T0386134

CHROMATIN

Structure
Dynamics
Regulation

CHAPMAN & HALL/CRC
Mathematical and Computational Biology Series

Aims and scope:
This series aims to capture new developments and summarize what is known over the entire spectrum of mathematical and computational biology and medicine. It seeks to encourage the integration of mathematical, statistical, and computational methods into biology by publishing a broad range of textbooks, reference works, and handbooks. The titles included in the series are meant to appeal to students, researchers, and professionals in the mathematical, statistical and computational sciences, fundamental biology and bioengineering, as well as interdisciplinary researchers involved in the field. The inclusion of concrete examples and applications, and programming techniques and examples, is highly encouraged.

Series Editors

Proposals for the series should be submitted to one of the series editors above or directly to:
CRC Press, Taylor & Francis Group
3 Park Square, Milton Park
Abingdon, Oxfordshire OX14 4RN
UK

Published Titles

An Introduction to Systems Biology: Design Principles of Biological Circuits
Uri Alon

Glycome Informatics: Methods and Applications
Kiyoko F. Aoki-Kinoshita

Computational Systems Biology of Cancer
Emmanuel Barillot, Laurence Calzone, Philippe Hupé, Jean-Philippe Vert, and Andrei Zinovyev

Python for Bioinformatics
Sebastian Bassi

Quantitative Biology: From Molecular to Cellular Systems
Sebastian Bassi

Methods in Medical Informatics: Fundamentals of Healthcare Programming in Perl, Python, and Ruby
Jules J. Berman

Chromatin: Structure, Dynamics, Regulation
Ralf Blossey

Computational Biology: A Statistical Mechanics Perspective
Ralf Blossey

Game-Theoretical Models in Biology
Mark Broom and Jan Rychtář

Computational and Visualization Techniques for Structural Bioinformatics Using Chimera
Forbes J. Burkowski

Structural Bioinformatics: An Algorithmic Approach
Forbes J. Burkowski

Spatial Ecology
Stephen Cantrell, Chris Cosner, and Shigui Ruan

Cell Mechanics: From Single Scale-Based Models to Multiscale Modeling
Arnaud Chauvière, Luigi Preziosi, and Claude Verdier

Bayesian Phylogenetics: Methods, Algorithms, and Applications
Ming-Hui Chen, Lynn Kuo, and Paul O. Lewis

Statistical Methods for QTL Mapping
Zehua Chen

An Introduction to Physical Oncology: How Mechanistic Mathematical Modeling Can Improve Cancer Therapy Outcomes
Vittorio Cristini, Eugene J. Koay, and Zhihui Wang

Normal Mode Analysis: Theory and Applications to Biological and Chemical Systems
Qiang Cui and Ivet Bahar

Kinetic Modelling in Systems Biology
Oleg Demin and Igor Goryanin

Data Analysis Tools for DNA Microarrays
Sorin Draghici

Statistics and Data Analysis for Microarrays Using R and Bioconductor, Second Edition
Sorin Drăghici

Computational Neuroscience: A Comprehensive Approach
Jianfeng Feng

Biological Sequence Analysis Using the SeqAn C++ Library
Andreas Gogol-Döring and Knut Reinert

Gene Expression Studies Using Affymetrix Microarrays
Hinrich Göhlmann and Willem Talloen

Handbook of Hidden Markov Models in Bioinformatics
Martin Gollery

Meta-analysis and Combining Information in Genetics and Genomics
Rudy Guerra and Darlene R. Goldstein

Differential Equations and Mathematical Biology, Second Edition
D.S. Jones, M.J. Plank, and B.D. Sleeman

Knowledge Discovery in Proteomics
Igor Jurisica and Dennis Wigle

Introduction to Proteins: Structure, Function, and Motion
Amit Kessel and Nir Ben-Tal

Published Titles (continued)

RNA-seq Data Analysis: A Practical Approach
Eija Korpelainen, Jarno Tuimala, Panu Somervuo, Mikael Huss, and Garry Wong

Introduction to Mathematical Oncology
Yang Kuang, John D. Nagy, and Steffen E. Eikenberry

Biological Computation
Ehud Lamm and Ron Unger

Optimal Control Applied to Biological Models
Suzanne Lenhart and John T. Workman

Clustering in Bioinformatics and Drug Discovery
John D. MacCuish and Norah E. MacCuish

Spatiotemporal Patterns in Ecology and Epidemiology: Theory, Models, and Simulation
Horst Malchow, Sergei V. Petrovskii, and Ezio Venturino

Stochastic Dynamics for Systems Biology
Christian Mazza and Michel Benaïm

Statistical Modeling and Machine Learning for Molecular Biology
Alan M. Moses

Engineering Genetic Circuits
Chris J. Myers

Pattern Discovery in Bioinformatics: Theory & Algorithms
Laxmi Parida

Exactly Solvable Models of Biological Invasion
Sergei V. Petrovskii and Bai-Lian Li

Computational Hydrodynamics of Capsules and Biological Cells
C. Pozrikidis

Modeling and Simulation of Capsules and Biological Cells
C. Pozrikidis

Cancer Modelling and Simulation
Luigi Preziosi

Introduction to Bio-Ontologies
Peter N. Robinson and Sebastian Bauer

Dynamics of Biological Systems
Michael Small

Genome Annotation
Jung Soh, Paul M.K. Gordon, and Christoph W. Sensen

Niche Modeling: Predictions from Statistical Distributions
David Stockwell

Algorithms in Bioinformatics: A Practical Introduction
Wing-Kin Sung

Introduction to Bioinformatics
Anna Tramontano

The Ten Most Wanted Solutions in Protein Bioinformatics
Anna Tramontano

Combinatorial Pattern Matching Algorithms in Computational Biology Using Perl and R
Gabriel Valiente

Managing Your Biological Data with Python
Allegra Via, Kristian Rother, and Anna Tramontano

Cancer Systems Biology
Edwin Wang

Stochastic Modelling for Systems Biology, Second Edition
Darren J. Wilkinson

Big Data Analysis for Bioinformatics and Biomedical Discoveries
Shui Qing Ye

Bioinformatics: A Practical Approach
Shui Qing Ye

Introduction to Computational Proteomics
Golan Yona

Chapman & Hall/CRC Mathematical and Computational Biology Series

CHROMATIN

Structure
Dynamics
Regulation

Ralf Blossey

CRC Press is an imprint of the
Taylor & Francis Group, an **informa** business
A CHAPMAN & HALL BOOK

CRC Press
Taylor & Francis Group
6000 Broken Sound Parkway NW, Suite 300
Boca Raton, FL 33487-2742

Printed on acid-free paper

International Standard Book Number-13: 978-1-4987-2938-3 (Hardback)

Visit the Taylor & Francis Web site at
http://www.taylorandfrancis.com

and the CRC Press Web site at
http://www.crcpress.com

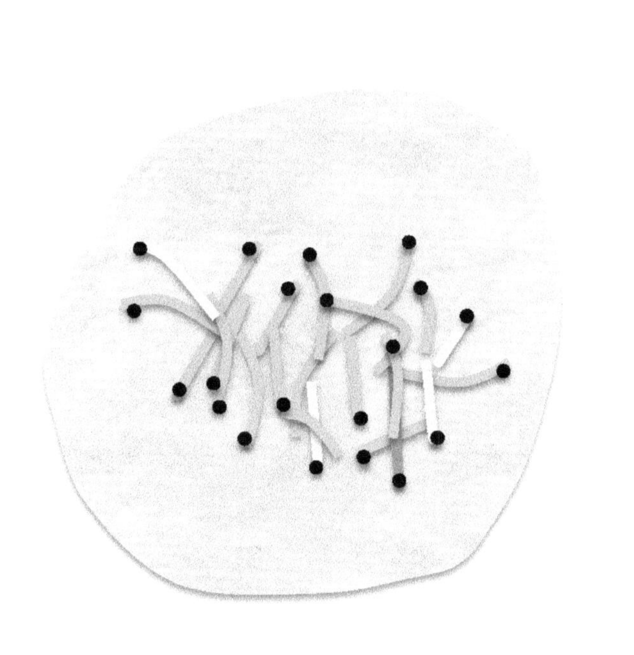

Contents

List of Figures

List of Tables

Theorie ist bei mir immer nachträglich.

Max Frisch

Preface

Chromatin, the genome-containing complex in the nucleus, is the central player in eukaryote biology. It is a huge macromolecular complex made up from DNA, histone proteins, and many other protein and RNA components. It contains the basic genetic information coded in DNA, but also all machinery to read it. In this analogy drawn with information technology, chromatin is both the software saved on the disk (DNA), but also all the electronics that go along with it to let the software run.

Understanding chromatin is a complicated task and meanwhile scientists with different backgrounds are active in trying to disentangle its properties. Biologists of different breeds have been joined by physicists and computational biologists. The purpose of this book is to provide a quick entry for a newcomer to the field who comes to the topic with a more "computational" background. I have attempted to present what is currently established about chromatin as far as its basic structure and involvement in gene regulation is concerned. Very much about chromatin is, however, although already qualitatively and quantitatively described, not yet really understood, and I came across a lot of debated topics when reading the literature. I have therefore avoided tracing these discussions here. Another difficulty for a newcomer to the field is that, common to biology, many chromatin-related topics are discussed in specific genetic or even medical contexts, which to the uninitiated hides the forest for the trees. This book is more about some of the trees. On the other side, developing a however "general" theory of chromatin is still a far way out. I try to reflect this in the motto I put at the beginning of the book, a quote from Swiss-German writer Max Frisch, which translates into English as *For me, theory always comes later.* In the case of chromatin, for all of us this is the case.

This book comes along with its subject dissected into three of its aspects, chromatin structure, dynamics and regulation. These topics are very much intertwined such that there is no natural or obvious way to separate them altogether. In the chapters of this book, the reader will therefore encounter these aspects in a sometimes mixed way. The material covered in the book is organized in six chapters, ranging from basic structural aspects to the dynamic and regulatory topics on different time and spatial scales.

Chromatin structure in this text is understood to be first of all the structure of the *chromatin fiber*, i.e., DNA and its organization by the histone proteins. Likewise, the notion of chromatin dynamics refers to the dynamic properties of this fiber and of the role the nucleosomes play, as well as the topoisomerases as one class of specific enzymes acting on DNA. A second set of chromatin-related enzymes, the so-called chromatin remodeling complexes, appear on stage in a chapter on histones and histone-acting enzymes. As this is a topic I worked on myself, it receives special attention here. Finally, this book interprets the notion of regulation in terms of the key biological function of chromatin, i.e., the switching on or off of *genes*, either temporarily or permanently.

This book results from several years of thinking about chromatin – mainly about its regulation – and the discussions I had with many colleagues. They are all thanked for the insights they shared with me, in particular my collaborators Guillaume Brysbaert, Fabrizio Cleri, Ana Maria Florescu, Marc Lensink, Helmut Schiessel, Raghav Singh and Yves Vandecan. I am also very grateful to numerous colleagues I met throughout the years and whose insights influenced my own path through this field. I thank particularly Alain Arneodo, Arndt Benecke (who introduced me to the topic), Bradley Cairns, Andrew Flaus, Ulrich Gerland, Jörg Langowski, Christophe Lavelle, Geeta Narlikar, Tom Owen-Hughes, Franklin Pugh, Karsten Rippe, Cédric Vaillant and Jean-Marc Victor and many others.

In the course of my interest in the field of chromatin I also enjoyed a number of discussions with Jon Widom. Jon's insights have set essential marks in the field; he is deeply missed by his colleagues for the unique combination of the profoundness of his scientific vision and his human side. I dedicate this book to his memory.

Ralf Blossey
Lille, May 2017

CHAPTER 1

DNA and the nucleosome

DNA structure

In our age of genomics, the biology of the cell is dominated by a DNA-centered view. In the early days of cellular biology this was not the case, as observations were limited to light microscopy. In the course of his studies of the transformations a eukaryotic cell undergoes – what we nowadays call the cell cycle – the biologist Walther Flemming observed the dynamical changes of the material in the nucleus of a cell and made drawings of the different stages of the cell cycle – like the sketch shown in the beginning of the book, which is inspired by the original drawings which can be found in the reprint (Flemming, 1965). Flemming called the complex structure in the nucleus *chromatin*, as it could be dyed and made visible in this way. In the course of the cell cycle, the nuclear material undergoes numerous visible shape changes, and, in particular in metaphase, condenses into the well-known form of X-shaped chromosomes.

In contrast to Flemming we now know that DNA is the substrate of these structures, and in order for it to undergo the structural changes a great number of different, very specific proteins intervene. In this book we try to understand the key features of this material as they are currently known. We will do this from the bottom up, hence start with DNA, and in the course of the chapters we will move up from the level of DNA to the chromatin complex in the cell.

The DNA molecule is a *nucleic acid* built from two single strands of nucleotides. Each of the strands is composed of a nitrogen-containing *nucleobase*, either cytosine (C) and thymine (T), the pyrimidines, and guanine (G) and adenine (A), the purines, and a monosaccharide sugar called deoxyribose and a phosphate group. The nucleobases form hydrogen bonds between corresponding Watson–Crick pairs, two for AT and three for CG, thereby linking the

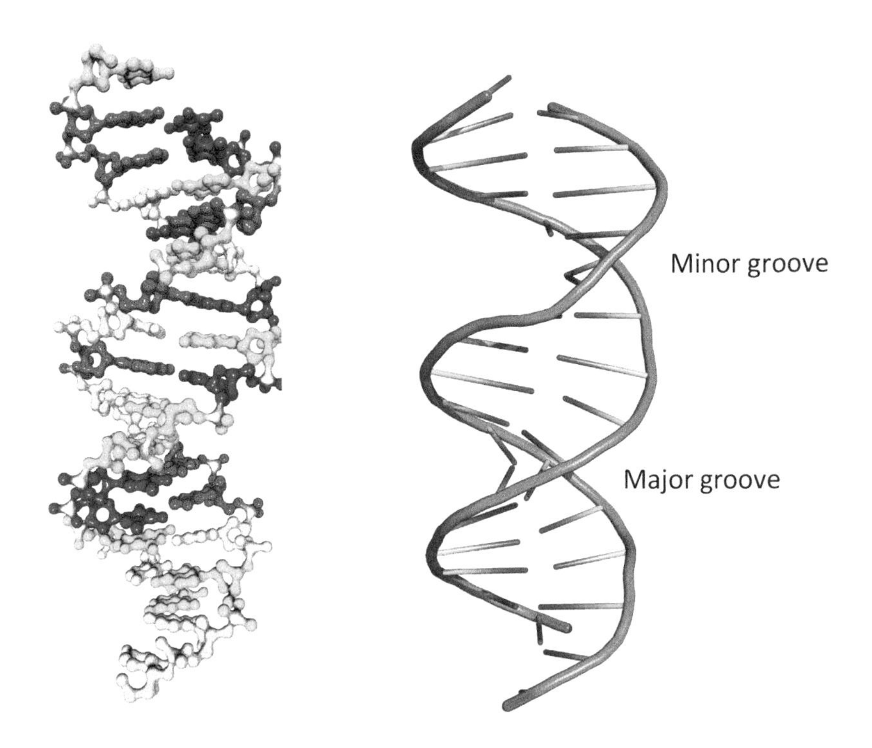

FIGURE 1.1 The molecular structure of DNA. Left: hyperball representation with color code as red = adenine, blue = thymine, green = guanine, yellow = cytosine. Realized with UnityMol 0.9.6. Right: cartoon representation of the same molecule, created with Pymol 1.7. Data taken from PDB entry 1HJC. Courtesy G. Brysbaert.

two strands into the double helix. The nucleobases are stacked on each other by covalent bonds between the sugar of one nucleotide and the phosphate of the next, which leads to an alternating sugar-phosphate backbone. Figure 1.1 displays an exemplary structure of the DNA double helix.

Both DNA strands wind around each other every 10.4 base pairs (bp), giving rise to the double-helical structure of DNA with its associated minor and major grooves, which are also visible in Figure 1.1. The DNA molecule can appear in several conformations which differ in the stacking of the base pairs; these are called the A-, B- and Z forms, whereby the A-form is a more compact structure than B-DNA, and the Z-form is left-winding. The B-DNA structure is the biologically most relevant one, and also the one originally observed in X-ray crystallography. Figure 1.2 shows a sketch of the original X-ray graph from

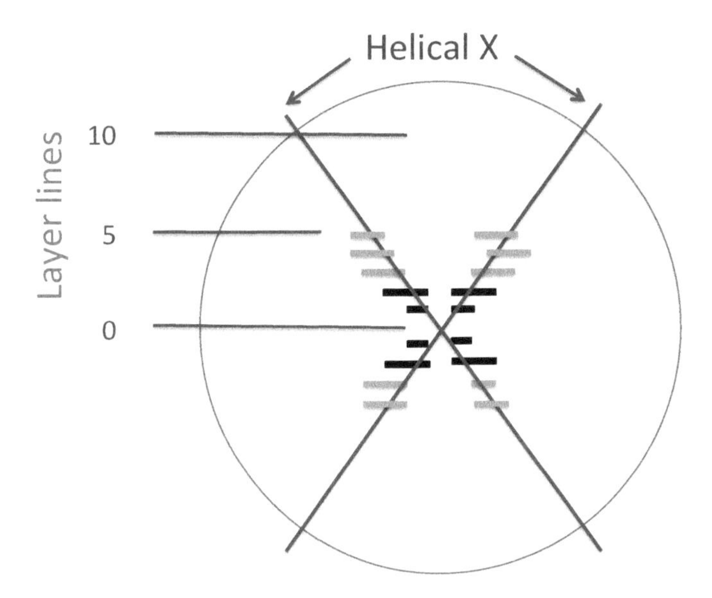

FIGURE 1.2 Sketch of the original X-ray diffraction pattern observed by Rosalind Franklin, which was the basis of the understanding of DNA double-helical structure. The lines corresponding to the stacked base pairs are indicated. Redrawn after P. Shing Ho and M. Carter (2011).

whose characteristic X-line shape the double helical nature of the molecule could be read off.

The DNA strand has an orientation: counting atomic positions around the ribose ring, the phosphate group is attached at the $5'$ position, while the oppositely lying $3'$ end is unmodified. In the double helix, one strands runs from $5'$ to $3'$, and the other in the opposite direction to allow for the pairing of the bases.

> **X-ray and TEM.** X-rays and TEM (transmission electron microscopy) are two methods for the imaging of molecular or material structures since both energetic photons and electrons can be diffracted at the molecular structure. X-ray has been the classic technique, developed for the studies of biomolecules mainly in the 1950s in Cambridge (UK). Electron microscopy is becoming an ever more important tool in structural biology in its low temperature version, cryo-EM; see also later in the book.

The X-ray diffraction pattern clearly offers an abstract view of the DNA molecule. Modern high-resolution imaging techniques can go further. A high-resolution image obtained with transmission electron microscopy (HRTEM) of the A-form of DNA was obtained only a few years ago (Marini et al., 2013).

The diameter of the DNA molecule is about 20 Å (slightly less for the Z-form, slightly larger for the A-form), but its length can reach up to meters. The sequence composition of DNA codes for most biological processes that a cell needs to perform, and in particular it contains the genes. A gene contains the information required for the production of a specific protein. In addition, a DNA sequence contains processing information of the genes through specific regulatory regions. In the context of gene regulation, we will see the working of these sequences in Chapter 3. For the moment we only mention a further distinction between the two main sequence types of DNA, the *exons* and *introns*. In higher organisms, eukaryotes — those organisms that have a nucleus in the cell — genes can be distributed over variable stretches of DNA, with non-coding segments interlaced. Exons comprise all DNA sequences that can be expressed (i.e., read and used for the production of proteins) while introns comprise non-expressed DNA (not in the form of proteins). We will come to some of the features displayed by exons and introns in Chapter 6 in the discussion of splicing.

In order to express genes, the double helix needs to be opened. This can be brought about by physical means through the increase of temperature, the pH of the solution, or the application of force. This process is called *DNA denaturation* or *DNA melting*, and has been the topic of numerous biophysical studies. As it is of less relevance for the properties of the chromatin fiber in the context we are interested in, we refer the interested reader to other introductory works (Blossey, 2006; Schiessel, 2014).

Finally, we need to mention that DNA is a charged molecule: DNA is negatively charged because of the phosphate group in the molecules' backbone, and it is thus sometimes called a "macroion" in a solvent. To compensate for the charge, DNA must be accompanied by counterions of opposite charge, or most typically by a solution with a given salt content. Monovalent and to an extent also divalent ions mostly compensate for the charges on DNA. Specific divalent ions, however, and in particular ions of higher valency, have important structural effects on the DNA molecule itself. These effects can even lead to attractive interactions even between nominally equally charged objects. Again, as for the problem of DNA melting, the question of the electric properties of DNA (and hence of the chromatin fiber) is not of primary concern in this book and we again point to other sources (Blossey, 2006; Schiessel, 2014) for those readers wishing to obtain an introduction to the problem. We will encounter the role of electrostatics in Chapter 4 in the discussion of some specific features of the nucleosome, the structure to which we now turn.

The nucleosome

The nucleosome is a protein–DNA complex in which the DNA winds around a protein core. The protein core of the nucleosome consists of an octameric complex, built from two copies each of the *histone proteins* H2A, H2B, H3 and H4. These together form a cylindrically shaped globular structure around which DNA is wrapped by about 147 base pairs (bps), i.e., about 1.7 times around the histone octamer. The histones are placed in the cited order in a top view of the nucleosome, in an anti-clockwise fashion starting in the upper right corner, as shown in Figure 1.3. A note on size: the height of the roughly cylindrical octamer complex measures up to 5.6 nm, its diameter to about 6.6 nm. Counting the DNA, the diameter of the nucleosome core particle (NCP) amounts to about 10.6 – 11 nm.

DNA is wrapped around the nucleosome in a left-handed manner, opposite to the winding of the DNA helix. This is often incorrectly represented in the literature, as C. Lavelle has pointed out very nicely (Lavelle, 2009). Figure 1.3 also shows a schematic drawing of the global orientation of DNA around the histone octamer, in side views from front and back. The nucleosome has a symmetry axis, called the *dyad axis*, which runs across the nucleosome midway through the octamer, with entering and exiting DNA to both the left and right sides. Positions on the nucleosomal DNA have been assigned by the symbol S (or SHL for *superhelical location*) and they are marked corresponding to the DNA helical pitch every 10.4 bp.

The formation of the nucleosome as a highly ordered protein–DNA complex is a regulated process which involves *histone chaperones*, i.e. specific protein assembly machines. There are several known histone chaperones with different specific functions. One of this class of molecules is the *nucleosome assembly protein*, NAP1. For this chaperone, which acts as a dimer, the kinetics of nucleosome formation has been studied in detail in *in vitro assays* by a measurement of the corresponding rate constants (Mazurkiewicz, Kepert and Rippe, 2006; Rippe, Mazurkiewicz and Kepper, 2008).

The build-up of the nucleosome complex proceeds in the following kinetic steps, involving first the transfer of two dimers of H3 and H4, and then the transfer of two dimers of H2A and H2B, with a hexasome intermediate state,

$$
\begin{aligned}
(\mathrm{H3}\cdot\mathrm{H4})_2 + \mathrm{DNA} \quad &\rightleftharpoons^{k_1}_{k_{-1}} \quad \mathrm{DNA}(\mathrm{H3}\cdot\mathrm{H4})_2 \\
\mathrm{DNA}(\mathrm{H3}\cdot\mathrm{H4})_2 + (\mathrm{H2A}\cdot\mathrm{H2B}) \quad &\rightleftharpoons^{k_2}_{k_{-2}} \\
&\qquad \mathrm{DNA}(\mathrm{H3}\cdot\mathrm{H4})_2(\mathrm{H2A}\cdot\mathrm{H2B}) \\
\mathrm{DNA}(\mathrm{H3}\cdot\mathrm{H4})_2(\mathrm{H2A}\cdot\mathrm{H2B}) + (\mathrm{H2A}\cdot\mathrm{H2B}) \quad &\rightleftharpoons^{k_3}_{k_{-3}} \\
&\qquad \mathrm{DNA}(\mathrm{H3}\cdot\mathrm{H4})_2(\mathrm{H2A}\cdot\mathrm{H2B})_2\,.
\end{aligned}
\tag{1.1}
$$

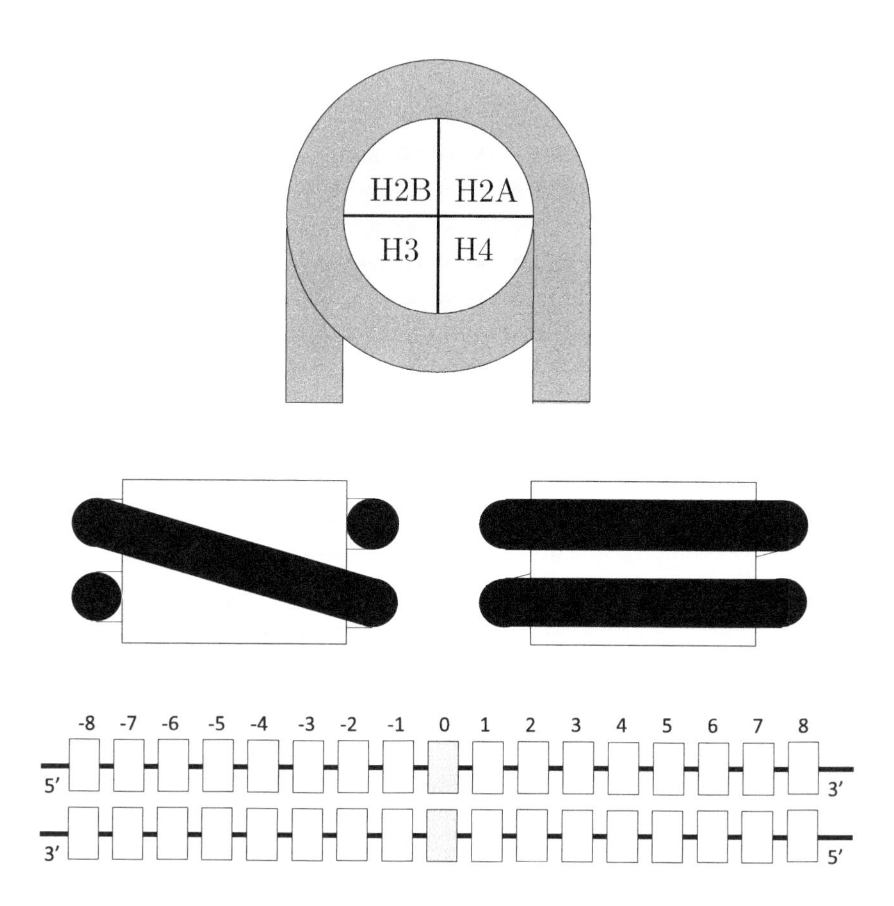

FIGURE 1.3 Schematic views of a nucleosome. Top: top view. Middle left: frontal view along the *dyad axis* with the DNA (black) entering at the bottom left, turning around at the back, mounting in a left helical manner to the upper left and completing the turn at the back. The DNA exits at the upper right corner. Middle right: the nucleosome from the back, showing the two parallel strands wrapping the histone. Bottom: nomenclature of DNA positions relative to the dyad axis at $S[0]$, shown in grey, toward the end of the nucleosomal DNA at $S[-8]$ to $S[+8]$, which have a spacing by 10.4 bp each. Positions are labeled from the $5'$ to the $3'$ ends for the two oppositely oriented DNA strands.

The concentration of NAP1 has been included in the definition of the reaction constants. The measured dissociation constants $K_{d,i} = k_{-i}/k_i$ for reactions $i = 1, 2, 3$ lie in the range

$$K_{d,1} \approx 150\,\text{nM}\,, \quad K_{d,2} \approx 5\,\text{nM}\,, \quad K_{d,3} \approx 190\,\text{nM}\,, \tag{1.2}$$

whereby these values depend on the length of the DNA used in the assay. The data above were reported for a short sequence of 146 bp, which just suffices to form a complete nucleosome core particle (NCP). The K_d-value of the last reaction went down by a factor of five for a 207-bp DNA. The observed values can also be put into relation with earlier in vivo measurements which yield comparable values if the differences in concentrations are accounted for. In a living cell, H3 and H4 bind to DNA on a sub-minute timescale, while H2A and H2B take 2–10 minutes.

There is another histone protein involved in the formation of the nucleosome: histone H1. In contrast to its relatives which form the globular nucleosomal octamer, histone H1 acts alone on the nucleosome by interacting with the entry-exit region of the DNA around the histone octamer. Early experiments by Pennings, Meerseman and Bradbury (1991; 1994) established that H1 inhibits nucleosome mobility. It therefore limits the access to nucleosomal DNA, but as it turns the nucleosome into a less dynamic structure, it also favors the formation of higher-order folding structures of the chromatin fiber.

One can include H1 in the nucleosome formation kinetics we discussed before,

$$\text{DNA(H3}\cdot\text{H4)}_2\text{(H2A}\cdot\text{H2B)}_2 + \text{H1} \underset{k_{-4}}{\overset{k_4}{\rightleftharpoons}} \text{DNA(H3}\cdot\text{H4)}_2\text{(H2A}\cdot\text{H2B)}_2\text{H1} \tag{1.3}$$

and one finds $K_{d,4}$ values which strongly depend on the NAP1 concentration, going down by about a factor of 10 when the NAP1_2-H1 concentration is reduced from 150 nM to 100 nM. This affinity increase is in accord with the dissociation constant of H1 in the absence of chaperones, which was found to be about 2 nM (Mazurkiewicz, Kepert and Rippe, 2006).

More recently, *hydroxyl footprinting* results in combination with molecular modeling could establish that the globular domain of histone H1, GH1, interacts with the minor groove located at the center of the nucleosome and contacts a 10-bp region of DNA localized symmetrically with respect to the dyad, and also with about one helical turn of DNA in the linker region of the nucleosome (Syed et al., 2010). H1 depletion affects the length of the *nucleosome linker region*, and as such again influences the formation of higher-order chromatin structure (Woodcock, Skoultchi and Fan, 2006).

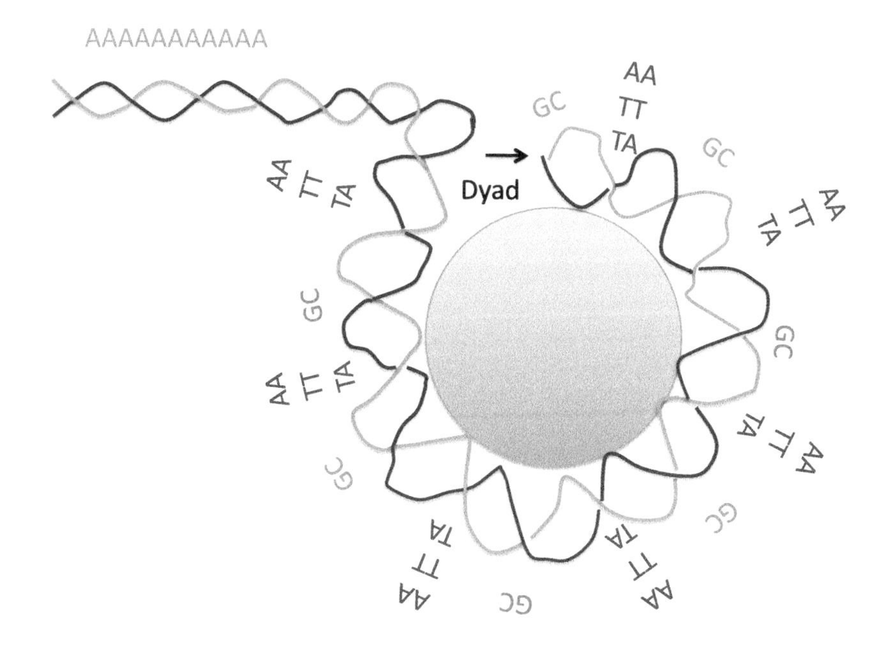

FIGURE 1.4 Dinucleotide positioning preferences indicated along the nucleosomal structure. Redrawn after Struhl and Segal (2013).

Hydroxyl footprinting. Hydroxyl footprinting is a technique in which a DNA sample is broken by added chemicals. If the DNA sample is exposed to the strand breaking agent once with and once without a protein bound to DNA, the results differ as the bound stretch of DNA is protected from the breaking agent. When comparing the bands of the broken DNA strands in gel electrophoresis, a gap will appear in the resulting gel ladder at the position of the protected sequence. Hydroxyl is advantageous with respect to other strand breaking agents as it has no sequence dependence. DNaseI is another such agent, a nuclease which was in fact used in experiments described below on the dynamics of the nucleosome. A nuclease is an enzyme which can cleave the phosphodiester bonds between the nucleotide subunits of nucleic acids. The size of DNaseI makes it easily hindered by a bound protein; however, it does have a sequence preference for its action.

Although the histone octamer does not have a direct sequence specificity, the positioning of a nucleosome is related to the base composition of the DNA.

This has to do with the bending of the DNA around the nucleosome, which depends on the sequence of base pair stacks, which is different for the different types of base-pair combinations. The precise mechanisms are still debated; a detailed discussion of the microscopic mechanisms can, e.g., be found in (Schiessel, 2014).

The key insight one needs to understand the position of nucleosomes along a given DNA sequence involves the role of *dinucleotide steps*, which refers to the stacking of two sets of base pairs (Drew and Travers, 1985). From the analysis of nucleosomal DNA, periodicities could be observed in specific steps: AA/TT, GC and TA. In the alignment of sequences of nucleosomal DNA, these steps appear in periodicities of ~10 bp, which is one helical turn of DNA. These occurrences have different phases, with the AA/TT and TA steps in phase, and the GC steps out of phase by 5 bp, half a helical turn of DNA.

These short sequence motifs can be linked to a higher flexibility in bending, which, however, is asymmetric, i.e., one set (AA/TT, TA) facilitates a bending of the DNA in one direction, and the other set (GC) in the opposite direction. In the analysis of mouse and chicken nucleosomes, the steps from the first set are located with highest probability at the positions of the nucleosome where the minor groove of the DNA faces inward, i.e. toward the center of the nucleosome. GC steps are thus facing outward, half the helical turn away; see Figure 1.4 following (Struhl and Segal, 2013) and Figure 1.5 (Segal et al., 2006). With these two Figures we already jump ahead of the discussion of nucleosome positioning, a topic to which we will turn in more detail shortly, but first we need to understand some of the dynamic properties of nucleosomes.

Nucleosome breathing and sliding

Even as a stably bound complex, the nucleosome is not a static object: it "breathes" or, more precisely, exposes the wrapped DNA to proteins for binding. This "site-exposure mechanism" has been detailed by J. Widom and collaborators in a series of papers (Polach and Widom, 1995; Anderson and Widom, 2000; Polach, Lowary and Widom, 2000).

In these experiments two types of reactions are compared in in vitro assays. Typically one could compare the binding kinetics of a protein to a DNA sequence that is in one situation occluded by the nucleosome and otherwise freely accessible on DNA. In this case one would find a difference between the two cases which can be expressed in terms of an apparent dissociation constant K_d^{app} given by

$$K_d^{app} = \frac{K_d^{nDNA}}{K_{eq}^c} \tag{1.4}$$

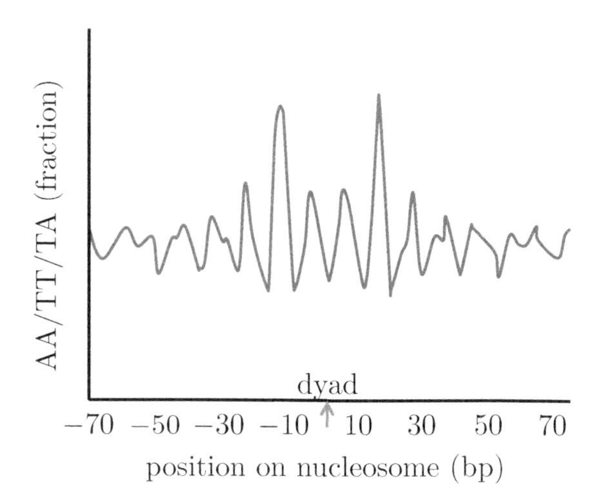

FIGURE 1.5 Typically observed fraction of AA/TT/TA dinucleotide steps at each position of center-aligned nucleosomal DNA sequences, showing their ~ 10-bp periodicity. Adapted from Segal et al. (2006).

where K_d^{nDNA} is the dissociation constant for naked DNA and

$$K_{eq}^{c} = e^{-\Delta G_c^0/RT} \tag{1.5}$$

where ΔG_c^0 is the free energy change associated with site exposure, which may be sequence dependent.

An even better and more sensitive way to conduct this experiment is to replace the protein by a nuclease digestion enzyme. The reactions to be compared are then given by

$$\mathrm{N} \rightleftharpoons_{k_{21}}^{k_{12}} \mathrm{S} + \mathrm{E} \rightleftharpoons_{k_{32}}^{k_{23}} \mathrm{ES} \rightarrow^{k_{34}} \mathrm{E} + \mathrm{P} \tag{1.6}$$

and

$$\mathrm{S} + \mathrm{E} \rightleftharpoons_{k_{32}}^{k_{23}} \mathrm{ES} \rightarrow^{k_{34}} \mathrm{E} + \mathrm{P} \tag{1.7}$$

in which the symbols denote the following: N is the wrapped nucleosomal state of the DNA, S a transient state with partial unwrapping, E is the nuclease and P the released product, a nucleosomal or DNA sequence of reduced length. The experiments fulfill a number of conditions on the rate constants, which are $K_{eq}^c \ll 1$; $[S] \ll K_m = (k_{32}+k_{34})/k_{23}$; $k_{21} \gg k_{23}(E)$ and $[E] \approx [E_0]$ where $[E_0]$ is the total concentration of the digestion enzyme. One can express K_{eq}^c

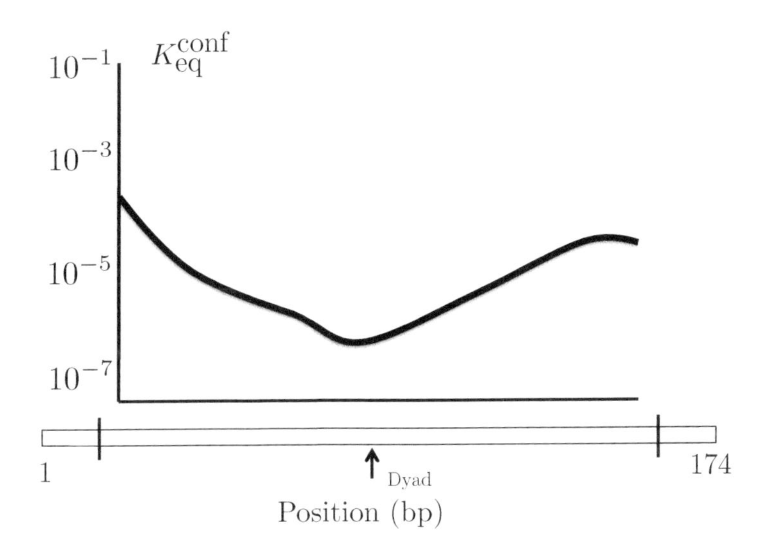

FIGURE 1.6 Qualitative sketch of the behavior of the measured equilibrium constants K_{eq}^c for the high-affinity nucleosome positioning sequence 601.2. Adapted from Anderson and Widom (2000).

as

$$K_{eq}^c = \frac{k_{obs}^N}{k_{obs}^{nDNA}} \frac{[E_0^{nDNA}]}{[E_0^N]} \tag{1.8}$$

where the rates k_{obs} are the first-order loss rates of the reactant in the two cases. The symbols, as before, relate to the experiment with "naked" DNA ($nDNA$) and in presence of the nucleosome (N).

Figure 1.6 summarizes the findings from these experiments for two positioning sequences which are either known or have been constructed to promote a strong positioning of nucleosomes. It is clearly visible that the equilibrium constants diminish downward from the edge of the sequence, and hence from the nucleosome border. The results show that the nucleosome must be dynamic as the DNA sequence it occludes is not entirely inaccessible.

A set of earlier experiments observed that nucleosomes bound to the sea urchin 5 S DNA sequence were not stably bound to a specific location on a 260 bp long sequence, but could be found 10 or 15 bp upstream. Is it possible that the nucleosome complex can actually move along a given sequence by thermal energy alone? Such presumed motion has been called *nucleosome sliding* and the two proposed mechanisms by which the complex can achieve this are

i) *Bulge or loop diffusion* (Schiessel et al., 2001). Bulge diffusion is a *polymer reptation process* around the nucleosome in which a DNA loop of length L^* detaches from the octamer core while at the same time some previously un-adsorbed linker DNA is pulled into the nucleosome. The pulled-in length ΔL is assumed to be 10 bp or multiples thereof. Solving the elasticity problem of the bent DNA around the nucleosome, one can calculate the diffusion constant associated with this process. The dependencies are given here only in scaling relations, omitting dimensional prefactors,

$$L^* \sim (\Delta L)^{1/3} \tag{1.9}$$

and

$$\Delta U \sim (\Delta L)^{1/3} \tag{1.10}$$

where ΔU is the loop formation energy. The loop diffusion constant is

$$D \sim (\Delta L)^{4/3} \exp(-\Delta U/k_B T)\,. \tag{1.11}$$

Based on these results, a quantitative estimate of the loop formation energy of a 10 base-pair loop then is at about $\Delta U \approx 23\, k_B T$, from which a diffusion constant $D \approx 10^{-17}$ cm^2/s follows. This is in accord with previous observations that the repositioning times are on the order of an hour.

ii) *Twist diffusion* (Kulić and Schiessel, 2003). The twist diffusion mechanism relies on the observation that nucleosome crystal structures have been found to contain 1 bp defects localized at a 10 bp stretch close to the dyad axis of the nucleosome. The diffusion of such a defect around the nucleosomal core can be described by employing models from crystal defect motion, which in the context of DNA describe the "creeping" of DNA around the histone core. The process therefore involves a twisting and stretching of DNA which can be modeled by an elastic energy of the form of harmonic springs

$$E_{elastic}(\{x_n\}) = \sum_k C \left(\frac{x_{k+1} - x_k}{b} - 1 \right)^2 \tag{1.12}$$

where the x_k are the positions of the k-th bp along the helical backbone, $k = 1, ..., n$, $b = 0.34$ nm is the equilibrium distance between the base pair stacks, and the elasticity constant $C \approx (70 - 100)k_B T$. The energy must be completed by the addition of an adsorption term describing the 14 contact points along the core (see Figure 1.4) and a sequence-dependent term. This term can be modeled as

$$E_{sd}(\ell) = (A/2)\cos(2\pi\ell/10 - \phi) \tag{1.13}$$

FIGURE 1.7 A caricature view of nucleosome positioning: boxes on a string.

where the amplitude A and the phase ϕ depend on the underlying sequence, and ℓ is the number of base pair steps. For the 5 S sequence mentioned before, the value of A is about $A \approx$ 5-6 kcal/mol. As in the case of loop diffusion, an analogous twist diffusion constant can be calculated which behaves as

$$D_{sd} \sim A \exp -(A + C/10)/k_B T \tag{1.14}$$

whose typical values are around $D_{sd} \sim 10^{-4}-10^{-5}\,(bp)^2/s$. On a DNA random sequence, this value can go up to $D_{sd} \sim 10^2\,(bp)^2/s$.

We have seen before that the position of nucleosomes is determined by DNA sequence — not in an entirely specific manner, but nevertheless there exist preferences for specific combinations of base stacks which favor a position pointing away from the histone octamer, and others that prefer to point inwards. These preferences are the basis for "optimized" positioning sequences for a single nucleosome (Lowary and Widom, 1998; Thåström et al., 1999).

Problem 1. Consider the effect of temperature on the bulge and twist diffusion constants.

But then — nucleosomes are not alone on the chromatin fiber. Considering just where a single nucleosome prefers to place itself freely is therefore not sufficient to know where a nucleosome will place itself under the constraint of the presence of several others, an issue to which we now turn.

Statistical positioning of nucleosomes

The first, and indeed very basic effect leading to a structuring of the chromatin fiber even without any specific sequence effects has been described by Kornberg and Stryer (1988); it is called *statistical nucleosome positioning.* If one considers that the nucleosomes sit on the chromatin fiber like particles

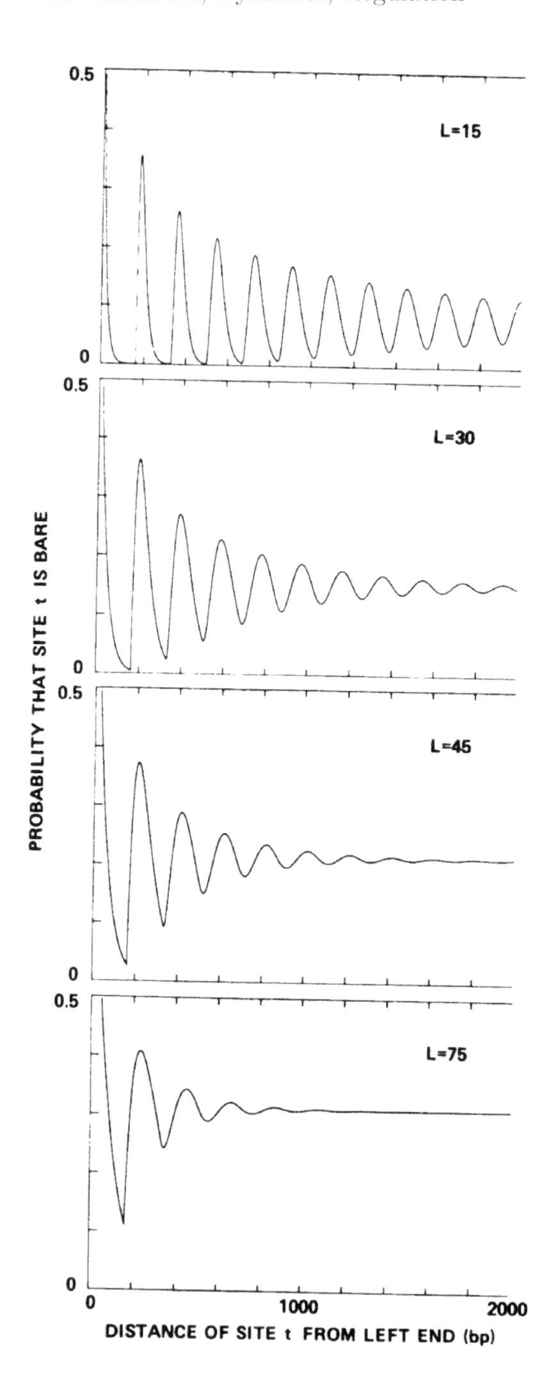

FIGURE 1.8 Probability of having an empty site on a lattice of length L on which particles of size d are placed. Reprinted with permission from Kornberg and Stryer (1988), © Oxford University Press.

on a one-dimensional string, as shown schematically in Figure 1.7, a simple positioning effect arises from the arrangement of the particles against a hard wall, which in the chromatin context is played by one very strongly positioned nucleosome, based on the bending properties of the DNA sequence. The resulting probability of finding a nucleosome next to this "wall" is shown in Figure 1.8. Similar behavior can be obtained in models from the statistical physics of fluids. Percus (1976) calculated the density of a classical fluid next to a wall which also shows these periodic variations. (The profile varies only in the direction perpendicular to the wall, as in the plane of the wall translational invariance can be assumed for the fluid.) Percus' result is actually employed in several of the methods used to predict nucleosome positions in genomic DNA, one of which we will discuss in Chapter 3.

Rather than using the apparatus of the statistical mechanics of fluids, we here follow the simple probabilistic treatment developed by Kornberg and Stryer which led them to compute the graphs in Figure 1.8. The starting point is the idea that the arrangement of nucleosomes along the DNA can be looked at as being an ensemble of "beads" (the nucleosomes) and empty stretches. We suppose that the width of a nucleosome is d (expressed in base pairs, hence a number), and we take L, also in base pairs, as the mean linker length, the length between two neighboring nucleosomes. The probability of selecting a bead from this ensemble is given by

$$v = \frac{1}{1+L} \tag{1.15}$$

and the probability of selecting an empty site is

$$u = \frac{L}{1+L}\,. \tag{1.16}$$

Note that the probability that a site on the DNA is filled is not given by u, but by $L/(d+L)$. The probability of selecting n beads and m empty sites is given by $v^n u^m$ so that the number of weighted ways (counting all possibilities to select them) to obtain the beads and intervals is given by the expression

$$w(g,n) = v^n u^m \frac{(n+m)!}{n!m!} \tag{1.17}$$

where $g = n + md$ is the number of base pairs. The sum of the $w(g,n)$ can run up to a maximal value of n, n_{max},

$$s(g) \equiv \sum_{n=0}^{n_{max}} w(g,n) \tag{1.18}$$

whose limiting value for $g \to \infty$ is given by $s_\infty = (1+L)/(d+L)$.

With these expressions we can now build probabilities for special cases. For example, the probability $p(a)$ that a site a positions away from the end of DNA of length ℓ is empty is given by

$$p(a) = \frac{s(a-1)w(1,0)s(\ell-a)}{s(\ell)} \tag{1.19}$$

which follows straight away from the fact that the presence of a single empty site along the DNA partitions the DNA into three pieces, one of length $a - 1$, which can contain beads, the empty site, and another stretch of DNA that can contain beads, of length $\ell - a$. If the DNA is considered as very long, $s(\ell - a)$ and $s(\ell)$ can be replaced by s_∞ and hence cancel out. The remaining result is plotted in Figure 1.8 as a function for two different values of L, taking as a value for $d = 166$ bp, a complete nucleosome including the linker part occluded by histone H1. The different values of L were motivated by experimental data of linker lengths for different organisms. The resulting curves show an oscillatory behavior with approximately the periodicity of the nucleosome repeat length $d + L$.

Problem 2. Use the Kornberg–Stryer model to compute the probability that two sites along the DNA are bare.

We leave the topic of nucleosome positioning at this point, and now proceed to another property of the chromatin fiber which we very briefly touched upon in the discussion of global chromatin structure. DNA is a polymer, and as such it has elastic properties. We have a look at those next.

References

Anderson J.D. and J. Widom
Sequence and position-dependence of the equilibrium accessibility of nucleosomal DNA target sites.
J. Mol. Biol. 296, 979-987 (2000)

Blossey R.
Computational Biology from a Statistical Mechanics Perspective.
Taylor & Francis (2006)

Drew H.R. and A. Travers
DNA bending and its relation to nucleosome positioning.
J. Mol. Biol. 186, 773-790 (1985)

Flemming W.
Contributions to the knowledge of the cell and its vital processes.
J. Cell Biol. 25, 1-69 (1965)

Kornberg R.D. and L. Stryer
Statistical distribution of nucleosomes: nonrandom locations by a stochastic mechanism.
Nucl. Acids Res. 16, 6677-6690 (1988)

Kulić I. and H. Schiessel
Chromatin dynamics: nucleosomes go mobile through twist defects.
Phys. Rev. Lett. 91, 148103 (2003)

Lavelle C.
Left is right, right is wrong.
EMBO Rep. 10, 1185-1186 (2009)

Lowary P.T. and J. Widom
New DNA sequence rules for high affinity binding to histone octamer and sequence-directed nucleosome positioning.
J. Mol. Biol.13, 19-42 (1998)

Marini M., A. Falqui, M. Moretti, T. Limongi, M. Allione, A. Genovese, S. Lopatin, L. Tirinato, G. Das, B. Torre, A. Giugni, F. Gentile, P. Candeloro and E. Di Fabrizio

The structure of DNA by direct imaging.
Sci. Adv. 1:e100734 (2015)

Mazurkiewicz J., J.F. Kepert and K. Rippe
On the mechanism of nucleosome assembly by histone chaperone NAP1.
J. Biol. Chemistry 281, 16462-16472 (2006)

Pennings S., G. Meersseman and E. Morton Bradbury
Mobility of positioned nucleosomes on 5 S rDNA.
J. Mol. Biol. 220, 101-110 (1991)

Pennings S., G. Meersseman and E. Morton Bradbury
Linker histones H1 and H5 prevent the mobility of positioned nucleosomes.
Proc. Natl. Acad. Sci. 91, 10275-10279 (1994)

Percus J.K.
Model for density variation at a fluid surface.
J. Stat. Phys. 15, 423-435 (1976)

Polach K.J. and J. Widom
Mechanism of protein access to specific DNA sequences in chromatin: a dynamic equilibrium model for gene regulation.
J. Mol. Biol. 254, 130-149 (1995)

Polach K.J., P.T. Lowery and J. Widom
Effects of core histone tail domains on the equilibrium constants for dynamic DNA site accessibility in nucleosomes.
J. Mol. Biol. 298, 211-223 (2000)

Rippe K., J. Mazurkiewicz and N. Kepper
Interactions of histones with DNA: nucleosome assembly, stability, dynamics and higher order structure.
DNA interactions with polymers and surfactants, Ch. 6, pp. 135-172, R. Dias and B. Lindman (eds.), John Wiley & Sons, Inc. (2008)

Schiessel H., J. Widom, R.F. Bruinsma and W.M. Gelbart
Polymer reptation and nucleosome repositioning.
Phys. Rev. Lett. 86, 4414-4417 (2001); Erratum: 88, 129902 (2002)

Schiessel H.
Biophysics for Beginners. A journey through the cell nucleus.
Pan Stanford Publishing (2014)

Segal E., Y. Fondufe-Mittendorf, L. Chen, A. Thåström, Y. Field, I.K. Moore, J.-P.Z. Wang and J. Widom
A genomic code for nucleosome positioning.
Nature 442, 772-778 (2006)

Shing Ho P. and M. Carter
DNA structure: alphabet soup for the cellular soul, DNA replication - current advances, Dr Herve Seligmann (Ed.), ISBN: 978-953-307-593-8, InTech, DOI: 10.5772/18536. (2011)

Struhl K. and E. Segal
Determinants of nucleosome positioning.
Nat. Struct. Mol. Biol. 20, 267-273 (2013)

Syed S.H., D. Goutte-Gattat, N. Becker, S. Meyer, M.S. Shukla, J.J. Hayes, R. Everaers, D. Angelov, J. Bednar and S. Dimitrov
Single-base resolution mapping of H1-nucleosome interactions and 3D organization of the nucleosome.
Proc. Natl. Acad. Sci. (USA) 107, 9620-9625 (2010)

Thåström A., P.T. Lowary, H.R. Widlund, H. Cao, M. Kubista and J. Widom
Sequence motifs and free energies of selected natural and non-natural nucleosome positioning DNA sequences.
J. Mol. Biol. 288, 213-229 (1999)

Woodcock C.L., A.I. Skoultchi and Y. Fan
Role of linker histone in chromatin structure and function: H1 stoichiometry and nucleosome repeat length.
Chromosome Res. 14, 17-25 (2006)

Further reading

Buning R. and J. van Noort
Single-pair FRET experiments on nucleosome conformational dynamics.
Biochemie 92, 1729-1740 (2010)

A review of experiments on nucleosome dynamics.

Luger K.
Nucleosomes: structure and function.
Encyclopedia of Life Sciences, 1-8 (2001)

Prinsen P. and H. Schiessel
Nucleosome stability and accessibility of its DNA to proteins.
Biochimie 92, 1722-1728 (2010)

This paper contains a recent and more detailed analysis of the Polach–Widom experiment.

CHAPTER 2

DNA elasticity and topology

DNA elasticity

In the previous chapter we encountered an example of a biochemical assay in the form of DNA fingerprinting. In this chapter we look at a particular type of *physical assay* which allows us to study the properties of biomolecules at a single-molecule level. This technique makes use of an important property of DNA – it is a flexible molecule. We therefore need to learn something about how to treat and characterize DNA as an elastic object. The first part of this chapter takes up elements from the discussion given in Schiessel (2014).

Let us consider the DNA as an elastic rod which can undergo bending and twisting, with corresponding energy costs. If we treat the DNA molecule in this way as a sort of macroscopic object and apply elasticity theory to it, we can write down the following energy expression, which is called the *worm-like chain model* (WLC)

$$H = \frac{1}{2}\int_0^L ds \left[A\left(\frac{1}{R(s)}\right)^2 + C\left(\frac{d\tau(s)}{ds}\right)^2 \right] \tag{2.1}$$

where s is the arc length of the rod, $R(s)$ the local radius of curvature along the rod which one can obtain by inscribing a local circle to the molecule, and $\tau(s)$ the local twist along the rod, as shown in Figure 2.1. The parameters A and C are the bending and twisting moduli, both quantities of dimension $[energy \times length]$, and L is the overall length of the rod.

In order to compute H for a given rod configuration, one can introduce a local triad of vectors to each point along the chain. We can simplify things here by considering the specific type of configurations we will be looking at to study DNA elasticity. Right from the start we will consider the case that is employed in the experimental single-molecule studies of DNA, in which a DNA

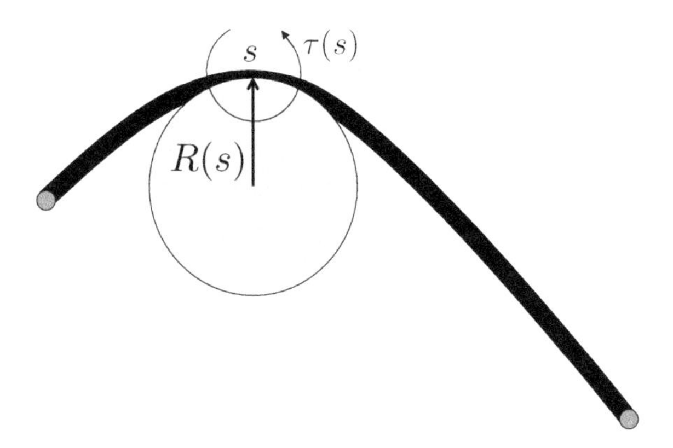

FIGURE 2.1 The definition of the local radius of curvature, $R(s)$, and the torsion, $\tau(s)$.

molecule is, e.g., fixed at a point on a substrate, and then dangles into the free space above the support. At the other end of the DNA molecule, a small bead is fixed. Figure 2.2 shows a sketch of this situation. The configuration of the dangling DNA, or even more precisely the motion of the bead at the extremity of the DNA, resembles the motion of the tip of a top in the gravitational field. This is a problem of classical mechanics which was first studied systematically by Leonhard Euler. In particular, he showed that the motion of the top can be described in terms of three angles that relate to the three possible motions in space. One can easily express the elastic energy in terms of these *Euler angles*, introduced in Figure 2.2, and find the expression of the WLC model

$$H = \int_0^L ds \left[\frac{A}{2}(\dot{\phi}^2 \sin^2\theta + \dot{\theta}^2) + \frac{C}{2}(\dot{\phi}\cos\theta + \dot{\psi}^2) - f\cos\theta \right] \tag{2.2}$$

where we have also added a force f to pull at the end of the chain.

In order to study the properties of the chain, we consider a simple case first. If we only allow for bending of the molecule in two dimensions, we can ignore any twist and also for the moment switch off the external force f. The energy H then simplifies to

$$H = \frac{A}{2}\int_0^L ds\, \dot{\theta}^2(s)\,. \tag{2.3}$$

We first want to understand what the chain does when it is kicked upon randomly by the molecules in which it is solvated, typically a salt solution. Remember that DNA is not a freely floating cylinder in air, but a molecular object surrounded by water molecules, plus the ions. We therefore must compare the elastic energy of the DNA molecule to the thermal energy k_BT which agitates the solvent environment around it.

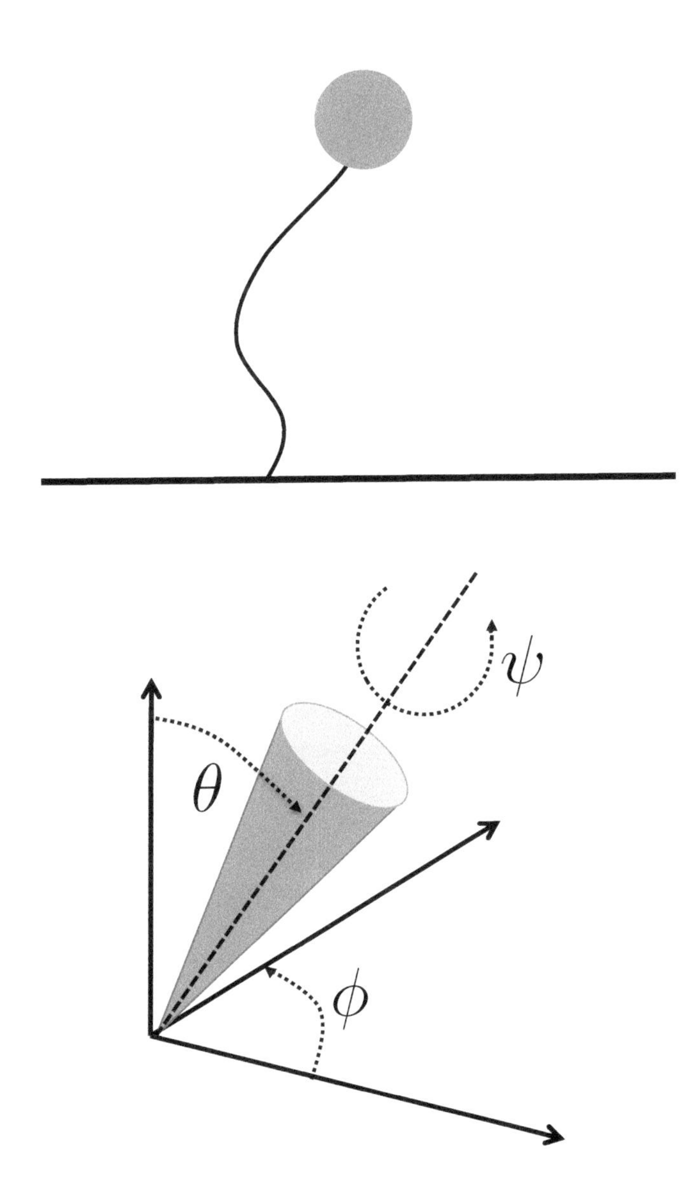

FIGURE 2.2 Top: a DNA molecule tethered to a surface. Bottom: a spinning top with the definition of the Euler angles.

The change in the local angle $\theta(s)$ along the chain is actually nothing but

$$\dot{\theta}(s) = \left(\frac{d\mathbf{t}}{ds}\right) \tag{2.4}$$

where $\mathbf{t}(s)$ is the local tangent to the curve introduced in Figure 2.3.

If we consider two points along the chain not too far away from each other at positions s and s', as shown in Figure 2.3, we can compute the correlation function of the tangent to the curve $\langle \mathbf{t}(s) \cdot \mathbf{t}(s') \rangle$ as an average with the Boltzmann factor $\exp -H/k_BT$, which yields the expression

$$\langle \mathbf{t}(s) \cdot \mathbf{t}(s') \rangle = e^{-|s-s'|/2L_p} \,. \tag{2.5}$$

In three dimensions, the factor 2 in the denominator of the exponential needs to be replaced by 1. In this formula, we have introduced the quantity L_p, which relates to the bending modulus of the polymer chain as

$$L_p \equiv \frac{A}{k_BT} \,, \tag{2.6}$$

i.e., it is the ratio of the bending modulus to the thermal energy. The result means that, for distances $x \equiv |s - s'| \gg L_p$, the chain has essentially "forgotten" about its orientation, as the tangent vector correlation has become exponentially small. The length L_p is called the *persistence length* of DNA and its typical value for double-stranded DNA at room temperature turns out to be about 50 nm. With the size of a base pair of 0.33 nm, this length corresponds to about 150 bp of sequence. Not only do the ends of the DNA molecule not know anything about what is happening at the other end, even two points along the chain separated by merely 200 bp are essentially ignorant of each other.

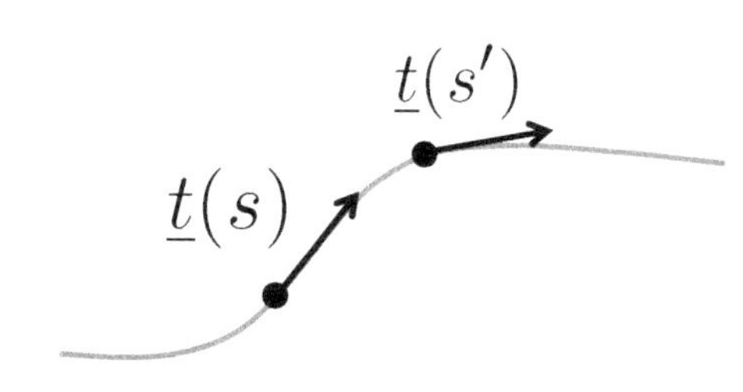

FIGURE 2.3 Two tangent vectors $\mathbf{t}$ to the curve describing the DNA molecule.

Next we consider the case when DNA is put under tension by pulling with a force f at the bead. In the limit that we stretch the chain almost straight, we can assume $\theta(s) \ll 1$, and the WLC expression simplifies to (we again assume two dimensions and neglect the twist term)

$$H = \int_0^L ds \left[\frac{A}{2}\dot{\theta}^2 + \frac{f}{2}\theta^2\right] - fL \tag{2.7}$$

where we have used $f \cos\theta \approx f(1-(1/2)\theta^2)$. The free energy is now a quadratic form so that when we compute the variation of the free energy with respect to θ by the derivative $\delta H/\delta\theta = 0$ this will lead to a linear equation. This linear equation can then be treated by expanding the angle θ in Fourier modes

$$\theta(s) = \sum_{n=-\infty}^{n=+\infty} \widehat{\theta}_n e^{-2\pi i n s/L} \tag{2.8}$$

which describe all the possible wiggly configurations that the molecule can display. If we finally impose periodic boundary conditions, $\theta(0) = \theta(L)$ (chosen for technical simplicity), we can write the energy H as

$$H = \sum_n \left(\frac{2\pi^2 A}{L} n^2 + \frac{fL}{2} \right) |\widehat{\theta}_n|^2 - fL\,. \tag{2.9}$$

The mean-squared amplitude of a mode with index n can be estimated as

$$\langle |\widehat{\theta}_n|^2 \rangle = \frac{k_B T}{\frac{4\pi^2 A}{L} n^2 + fL} \tag{2.10}$$

when one assumes an equipartition of energy on all modes indexed by n.

This expression allows us to get an idea of what the force f does. Obviously, pulling on the molecule will straighten it out and thus suppress some of the wiggly configurations. Mathematically this follows from expression (2.10) by comparing the two terms in the denominator,

$$4\pi^2 A < \left(\frac{L}{n} \right)^2 f\,. \tag{2.11}$$

In this expression, an increase of the force f will stretch the DNA and suppress the modes of length L/n obeying

$$\frac{L}{n} > 2\pi\lambda\,, \tag{2.12}$$

with the definition of the force-induced correlation length λ through

$$\lambda \equiv \sqrt{\frac{A}{f}}\,. \tag{2.13}$$

We can also calculate the average extension of the chain, which is given by

$$\langle \Delta z \rangle = \int_0^L ds \langle \cos\theta(s) \rangle \approx \int_0^L ds \left[1 - \frac{1}{2} \langle \theta^2(s) \rangle \right] = L \left(1 - \frac{k_B T}{4\sqrt{Af}} \right) \tag{2.14}$$

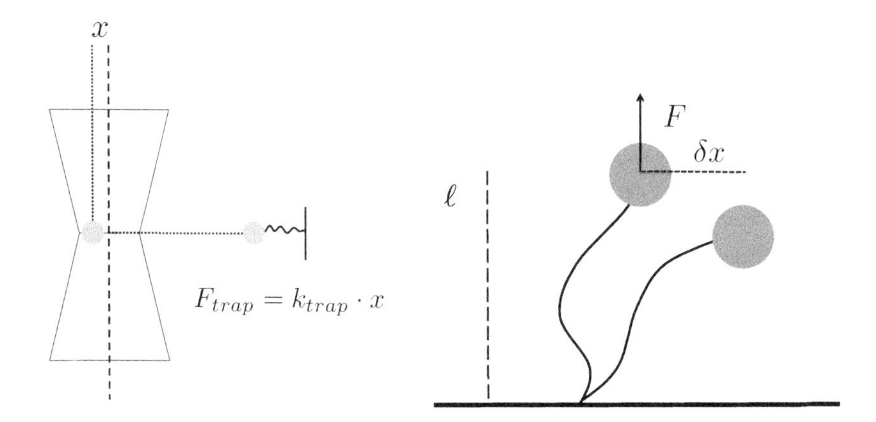

FIGURE 2.4 Left: An optical trap. Right: Measuring tethered DNA.

where in the calculation the sum over Fourier modes has been replaced by an integral over continuous n. Solving for the force f, we obtain

$$f = \frac{k_B T}{16 L_p} \frac{1}{\left(1 - \frac{\langle \Delta z \rangle}{L}\right)^2} . \tag{2.15}$$

In three dimensions, the value 16 in the denominator needs to be replaced by 4 in Eq. (2.15).

Magnetic and optical tweezers

How can this stretching experiment be done in reality? The two key ingredients are the fixing of the DNA to a substrate and, at the other end of the chain, to fix an object which can be manipulated by force. The two main methods are *magnetic* and *optical tweezers* in which a small bead is fixed to the DNA, which either responds to a magnetic field gradient, which is the case if a superparamagnetic bead is used, or to a focused laser beam in the optical case, in which electrical field gradients operate. The particle is pulled toward the strongest field, which is the position of the center of the beam. Figure 2.4 (left) sketches the optical trap setup.

How are the forces and elongations measured? To determine the force on the bead, one needs to balance the exerted force with the thermal fluctuations acting on the molecule. Figure 2.4 (right) shows the typical geometry, again only in two dimensions for simplicity. Any deviation δx of the chain from the center position is related to the restoring force via

$$F_x = \frac{F}{\ell} \delta x . \tag{2.16}$$

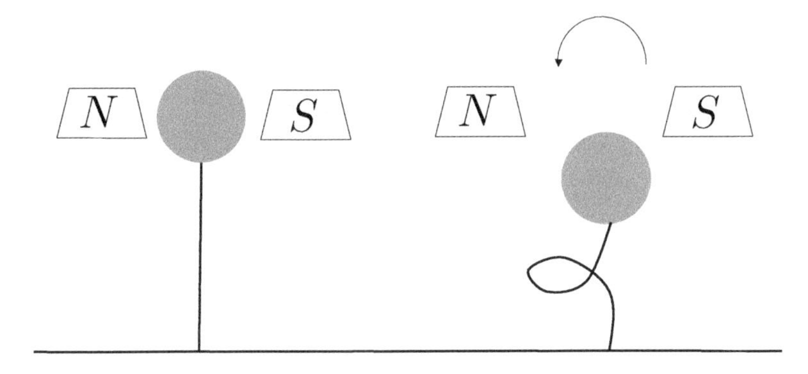

FIGURE 2.5 Sketch of a supercoiling experiment with magnetic tweezers. The setup is illuminated from above and the bead position is found from the diffraction patterns it produces, which can be detected by a CCD camera with a precision of a few nanometers. The precision in the vertical direction is less and a comparison to calibration images is needed.

By equipartition of energy we can estimate the average (potential) energy associated with this deviation as

$$\langle E \rangle = \frac{1}{2}\frac{F}{\ell}\langle \delta x^2 \rangle = \frac{1}{2}k_B T \tag{2.17}$$

and then obtain the expression

$$F = \frac{\ell k_B T}{\langle \delta x^2 \rangle}, \tag{2.18}$$

which links the exerted force F to the extension of the molecule ℓ and the mean-squared deviation from the center position, $\langle \delta x^2 \rangle$.

Knowing now that we can exert and measure a force on a tethered DNA molecule, e.g., in the magnetic tweezer setup, we can start to go beyond simple elastic effects; see Figure 2.5. One can turn the magnet and lock the bead into this turning motion. Depending on the imposed winding direction, we can in this way either over- or underwind the DNA. If we wind, we start to shorten the DNA extension and at a certain threshold, the strand will start to buckle and bring about a *plectoneme*, a supercoil.

We can now apply this method to DNA. Figure 2.6 shows the force-extension curve measured on λ-phage DNA (for more on this organism, see Chapter 3) with magnetic beads. The inset explains the comparison to different models, in particular the WLC model. An extrapolated formula for the WLC model reads as

$$f = \frac{k_B T}{4L_p} \frac{1}{\left(1 - \frac{\langle \Delta z \rangle}{L}\right)^2} + \frac{\langle \Delta z \rangle}{L} - \frac{1}{4} \tag{2.19}$$

and differs only slightly from Eq. (2.15). The curves show excellent agreement with the WLC model over three orders of magnitude. Figure 2.7 shows a result from the earliest supercoiling experiments with magnetic tweezers, dating back about 20 years. In order to interpret these latter curves, we need to introduce more concepts.

These concepts refer to the topology of the DNA. In the experiment discussed above both ends of the DNA are kept fixed, hence, e.g., a full rotation of the bead will change the degree of winding of the two chains around each other. This is captured by a topological constant, the *linking number* Lk. It is defined as the sum of two contributions, *twist* and *writhe*,

$$Lk = Tw + Wr\,. \tag{2.20}$$

Twist Tw reflects the helical winding of the strands around each other, and any DNA has, for a given length, a basic twist Tw_0, which is the number of helical turns of the two strands. Writhe Wr, on the other hand, describes the winding of the central axis of the DNA molecule around itself in space. If there is no writhe, say initially in the experiment, one has $Lk = Lk_0 = Tw_0$, where $Lk_0 = L/h$, where $h \approx 3.6$ nm is the contour length of one helical turn. The relative change in linking number is given by the degree of supercoiling

$$\sigma = \frac{Lk - Lk_0}{LK_0} = \frac{\Delta Lk}{LK_0}\,, \tag{2.21}$$

as depicted in the experimental curves of Figure 2.7. These curves display several important features. At low degrees of supercoiling, the extension of DNA is very well described by the worm-like chain model, and for low forces, there is a symmetry between positive and negative supercoiling. For a higher degree of supercoiling, the curves still follow the wormlike chain behavior, however with smaller persistence lengths. The asymmetry between the cases of positive and negative supercoiling is particularly visible in the relative extension vs supercoiling plots, which show the effect of the chirality of DNA. This is easy to understand qualitatively: when the added windings increase the helicity of the DNA, this will reach a limit above which the chain will start to buckle. However, when the windings are subtracted, the double helix becomes more and more unwound. Twisting DNA is indeed a very important effect as the molecule is twisted to begin with due do its very structure. If one now adds more twist and overwinds the DNA, as DNA is a very long molecule, it will also bend back and overlay itself in space. Just one thing it cannot do — it cannot interpenetrate itself.

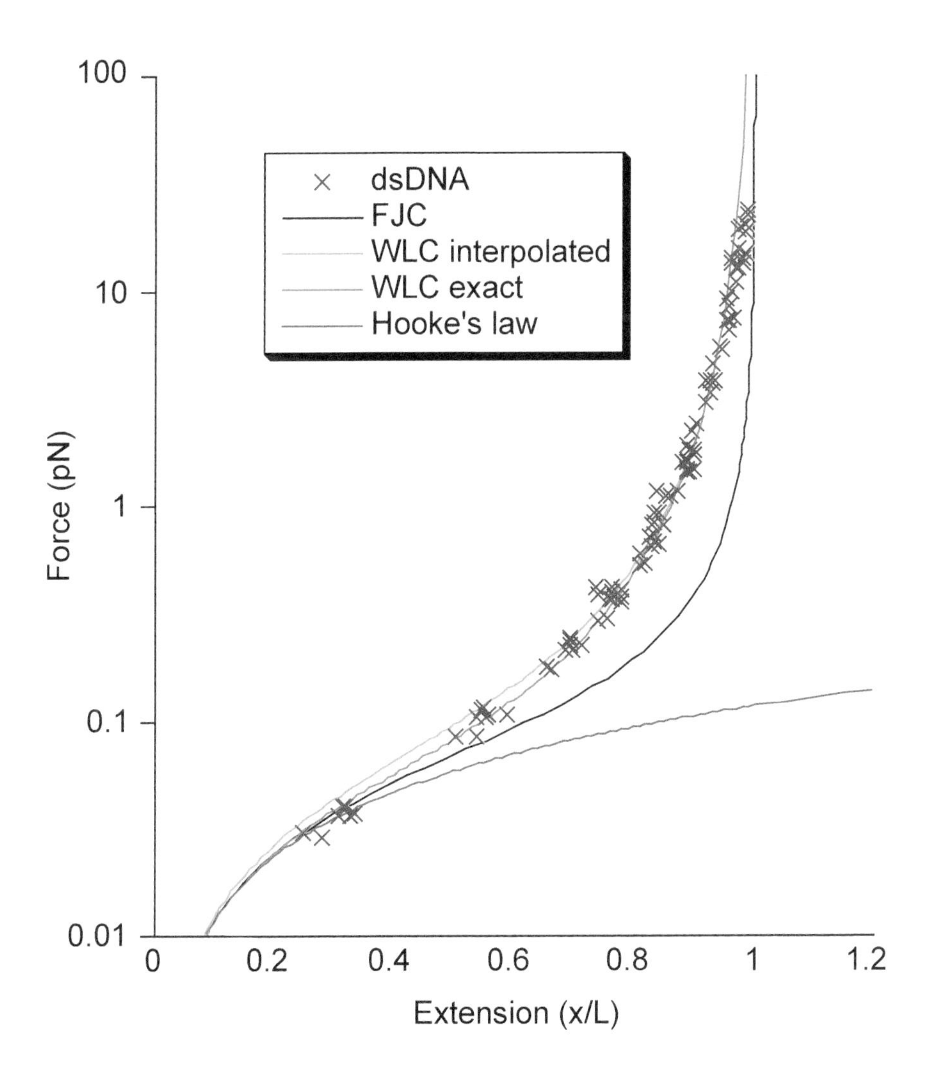

FIGURE 2.6 Force-extension curves for λ-phage double-stranded DNA attached to magnetic beads compared to a WLC model as well as other models. FJC: freely jointed chain model, an even simpler polymer model which represents the molecule as a random walk. Reprinted with permission from Bustamante et al. (2000) © Elsevier.

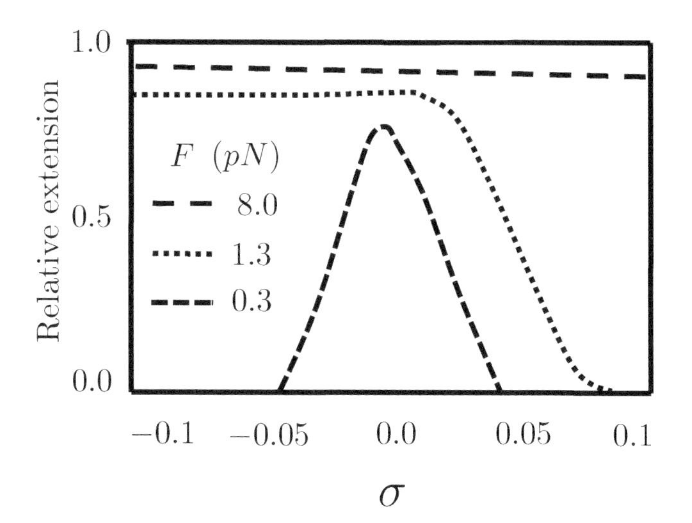

FIGURE 2.7 Relative extension vs. supercoiling σ, positive and negative. At higher degrees of supercoiling the curves become asymmetric. Qualitative drawing after Strick et al. (1996).

Adding now the twist contribution to the chain energy, we can write

$$\beta E_{twist} = \frac{C_p}{2L}\theta^2 \tag{2.22}$$

where θ is the twist angle along the double helix, and the parameter C_p is the bending stiffness of the WLC model, divided by the thermal energy, in analogy to the bending persistence length L_p. This is the *twist persistence length*. Without bending, one has $\theta = 2\pi\Delta Lk$, and hence

$$\sigma = \frac{\theta h}{2\pi L}\,. \tag{2.23}$$

Thermal fluctuations lead to twist angle fluctuations $\langle\theta^2\rangle = L/C_p$, with a value of $C_p \approx 100$ nm. The derivative

$$\tau \equiv \frac{\partial E_{twist}}{\partial\theta} = k_B T \frac{C_p}{L}\theta \tag{2.24}$$

is the torque or torsional stress present in DNA. Expressing this in terms of σ, we have

$$\tau = \frac{2\pi C_p}{h} k_B T \sigma \tag{2.25}$$

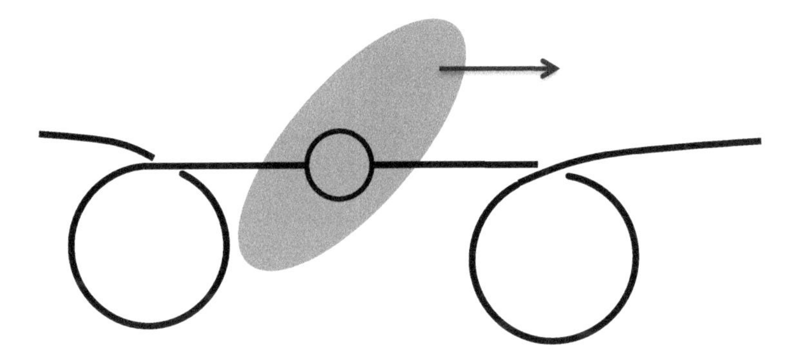

FIGURE 2.8 Sketch of supercoiling due to transcription by an RNA polymerase. Adapted from Mirkin (2001).

and we see that, similar to our reasoning above, τ becomes appreciable for DNA when $|\tau| \approx k_B T$, hence

$$|\sigma| \approx 0.005 \tag{2.26}$$

for a typical value of $2\pi C_p/h \approx 175$.

More on twist and writhe

Supercoiling is not just an experimental effect, it is a natural "problem" the cell faces in transcription (and this is true for both prokaryotes and eukaryotes). When an RNA polymerase complex has acquired access to a gene and starts the transcription process, it has to open the strands to read the DNA sequence in order to perform the copying process of the gene. Pulling on a piece of DNA and ripping the strands apart affects the coiling status of the molecule. This is shown in a sketch in Figure 2.8. The RNA polymerase is shown as a grey ellipse in the center of which a *transcription bubble* has opened. The direction of transcription is indicated by the arrow. The translocation forces the DNA to rotate around the polymerase so that negative and positive supercoiling is produced upstream and downstream of the polymerase. These supercoilings need to be relieved and the cell does this with enzymes called *topoisomerases*, which come in two types called Topo I and Topo II. Type I topoisomerases relieve the stress in the DNA by cutting a single strand, a

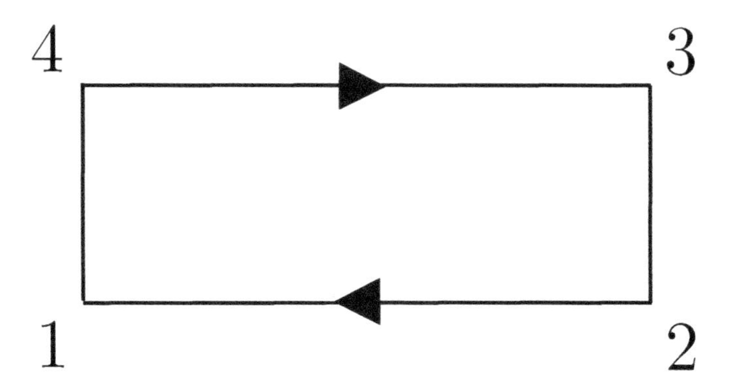

FIGURE 2.9 Ribbon representation of a DNA molecule. The orientations are derived from the oriented nature of the two strands making up the DNA molecule.

process which requires no ATP. Type II topoisomerases cut both strands, a process which consumes ATP.

Before discussing the properties of these enzymes in more detail, we first continue the discussion of the concepts of twist and writhe, and we do so for the particular case of a circular DNA. Twist Tw and writhe Wr are easier to visualize in a two-dimensional projection, which turns the rod-like DNA into a ribbon. In Figure 2.9 such a ribbon is sketched, with the long sides or edges of the ribbon being oriented, as they are supposed to represent the two antiparallel DNA strands. In the figure we have numbered the edges of the ribbon from 1 to 4 in a counter-clockwise fashion (the arrows are clockwise).

We can now close the ribbon to form a circular ribbon of DNA respecting the arrows. Such DNA molecules exist indeed in bacteria, where they are called *plasmids*. Further abstracting in our drawing we even leave out the ends of the ribbon and just represent the linked edges by oriented circles in Figure 2.10. The circles then appear as a disconnected graph. This DNA plasmid is neither twisted, nor do the circles overlap each other, so both Tw and Wr should be zero, hence $Lk = 0$.

Suppose now that we deform the ribbon by tweaking the right hand side toward us. The result is shown in Figure 2.11. The top edge runs between 4 and 3, but now from the left upper corner down to the right lower corner, and vice versa for the other edge. (We have left out the numbers, but their placement should be clear from Figure 2.10.) If we now want to complete the

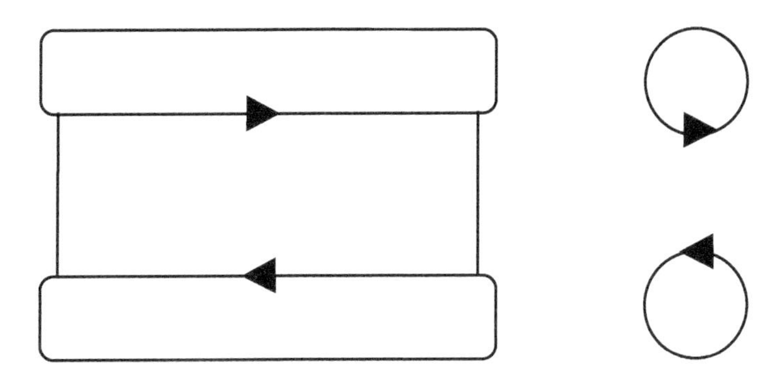

FIGURE 2.10 Closing the ribbon to a circle.

ribbon to a plasmid, in order to respect orientation of the edges we need to insert a second twisted piece of ribbon, in which now 1 and 4 have exchanged position from top to bottom. This piece results from a flipping over the first piece, which also reverts the orientation of the arrows. Now we can close the circles again and obtain a twisted plasmid. This can be seen even more easily in the circular abstraction of the twisted ribbon which is also shown in Figure 2.11.

In order to quantify the topology of this object, we focus on the crossings of the lines. There are four possible types, shown in Figure 2.12. The linking number can then be computed from the crossings of the lines via

$$Lk = \frac{n_1 + n_2 - (n_3 + n_4)}{2} . \tag{2.27}$$

For our example $n_3 = n_4 = -1$, $n_1 = n_2 = 0$, hence we have $Lk = +1$.

The twist of the circular ribbon is defined such that for a closure of the ribbon after its full rotation, i.e., by 360° , $Tw = +1$ if the twist is right-handed, or $Tw = -1$, if it is left-handed. Consistent with the above, in our case the ribbon is right-handed.

To complete this exercise we need to see what the writhe is. This can also be easily gathered from our simple case. In Figure 2.10, $Wr = 0$. If we close the ribbon to a circular plasmid-like form, it will change into three-dimensional space into a curve which has roughly the shape of an eight. In this way, the ribbon relaxes the strain that was imposed on it, and twist is exchanged for writhe. If we draw the relaxed curve in a projection to the plane as we do in Figure 2.13, we can compute the writhe of a molecule simply by

$$Wr = \sum_{i=crossings} (n_{1,i} + n_{2,i} - (n_{3,i} + n_{4,i})) \tag{2.28}$$

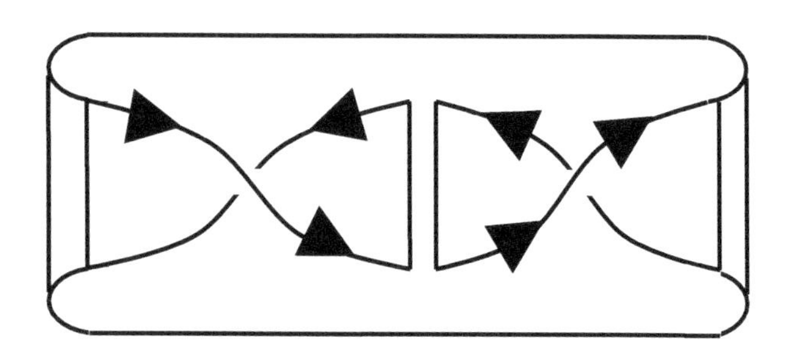

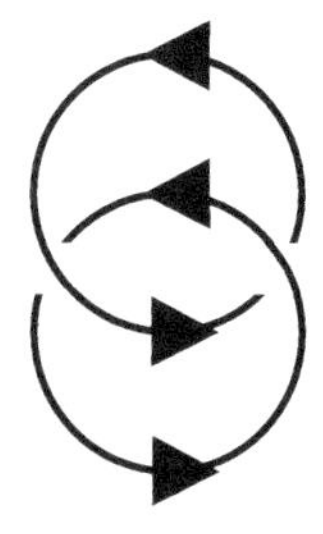

FIGURE 2.11 Closing the ribbon to a circle, but with a twist.

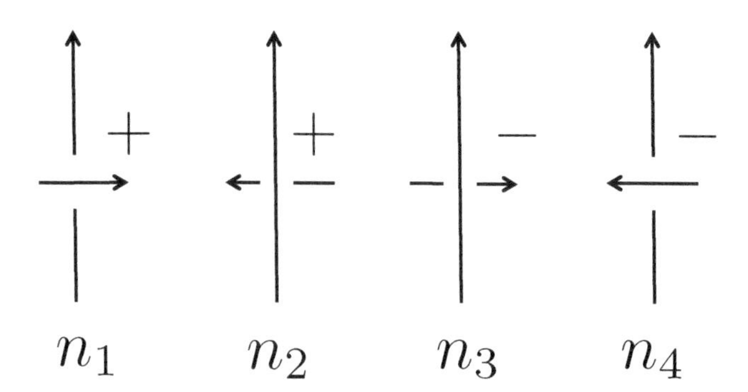

FIGURE 2.12 The four possibilities of strand encounters.

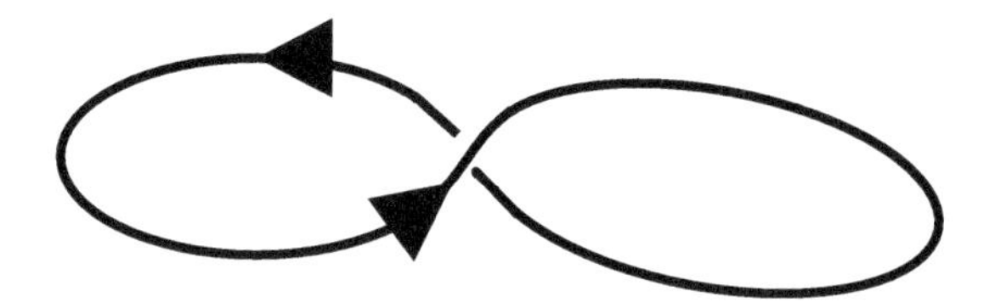

FIGURE 2.13 Relaxed ribbon, transforming twist into writhe. Only the centerline of the ribbon is shown in this sketch.

whereby the crossing numbers are defined as in Figure 2.12.

Problem 3. Perform the operations described in explaining Figures 2.10 and 2.11 with a paper strip.

Problem 4. Determine twist and writhe for the configurations of DNA minicircles a), b) and c) shown in Figure 2.14.

In this intuitive way we have gathered an understanding of these quantities, but the computation we showed is limited to "ideal" cases. It is important to note that the linking number Lk is a topological invariant (as in our example, where upon relaxation twist is removed at the cost of producing writhe). However, in our simple example only integer values arise, while in general both twist and writhe need not be integers. In the general case, the writhe is given by the line integral

$$Wr = \frac{1}{4\pi} \int ds \int ds' \frac{\mathbf{r}(s) - \mathbf{r}(s')}{|\mathbf{r}(s) - \mathbf{r}(s')|^3} \cdot \frac{d\mathbf{r}(\mathbf{s})}{ds} \times \frac{d\mathbf{r}(s')}{ds'}$$

where $\mathbf{r}(s)$ is the vector of the center line of the ribbon, parametrized by its arc length s. The linking number is given by a similar expression, the Gauss integral

$$Lk = \frac{1}{4\pi} \oint d\mathbf{r}_1 \oint d\mathbf{r_2} \frac{\mathbf{r}_1 - \mathbf{r}_2}{|\mathbf{r}_1 - \mathbf{r}_2|^3} \cdot d\mathbf{r}_1 \times d\mathbf{r}_2 \tag{2.29}$$

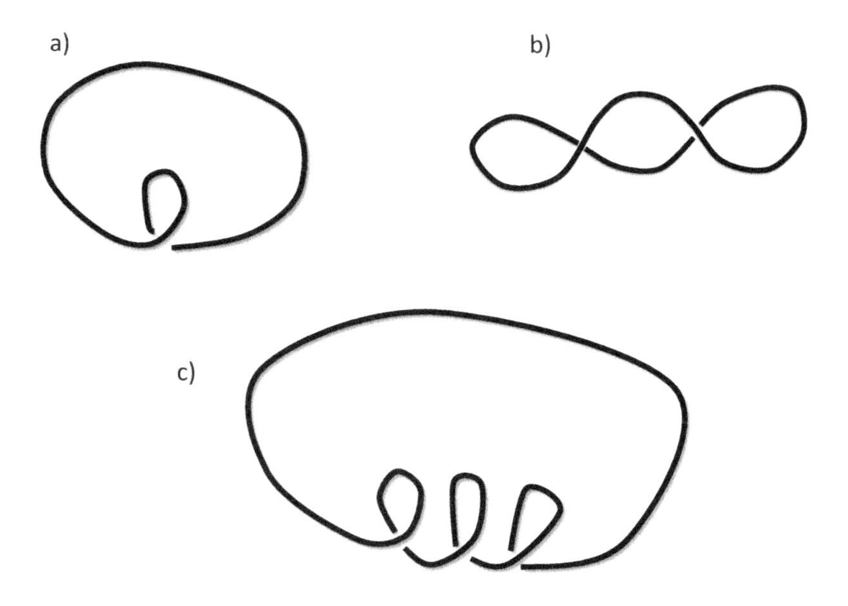

FIGURE 2.14 Twisted and writhed DNA minicircles.

where the vectors $\mathbf{r}_1$ and $\mathbf{r}_2$ describe each one of the two curves whose linking is considered.

Topoisomerase II

Topoisomerase I enzymes cut one of the strands such that the neighboring strand can pass through, and the cut strand is then resealed. Topoisomerase II enzymes are much more complex and interesting, due to the ATP dependence of their action.

Topo II action on DNA has been studied in a follow-up to the single-molecule experiments by magnetic tweezers in which supercoiling was induced by a rotation of the magnets. Figure 2.15 shows the same setup, but now with the addition of a topoisomerase II enzyme. In Figure 2.15b we see that the enzyme targets the coil induced by the rotating magnet and reduces the extension by an amount δ. Figure 2.15c shows how by consumption of ATP the action of the magnet is counteracted by the enzyme. This means that one can mechanically induce a supercoil, which is removed by the enzyme.

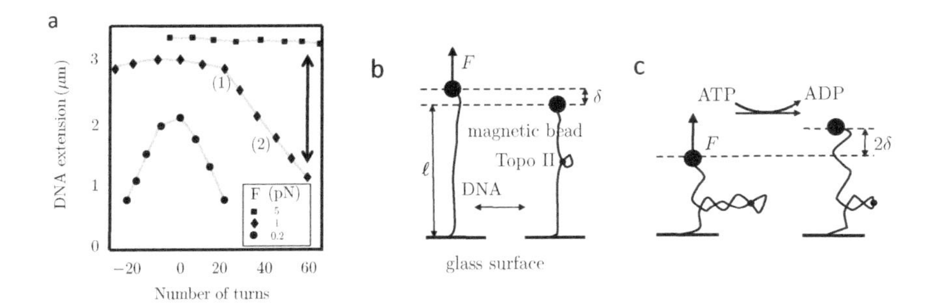

FIGURE 2.15 a) At low forces, the molecule buckles without denaturation; at intermediate forces, the DNA needs to be overwound by 25 turns before it starts to buckle at point (1), from where on it continues to buckle (2). For underwinding, denaturation occurs. b) Addition of Topo II reduces the extension by an amount δ. c) Under ATP, the DNA relaxes by an amount of 2δ as each enzymatic cycle releases two supercoils. Redrawn after Strick, Croquette and Bensimon (2000).

ATP. Adenosine triphosphate (ATP) is a nucleoside triphosphate, a small molecule used in cells as a coenzyme. It provides chemical energy within cells for metabolism. The standard amount of energy released from hydrolysis of ATP is

$$ATP + H_2O \to ADP + P_i \tag{2.30}$$

with the enthalphy $\Delta G^o = -30.5$ kJ/mol.

Two characteristic behaviors are sketched in the figure. In Figure 2.15b, it is shown that the enzyme targets the supercoil and "clamps" it, thereby reducing, even without ATP, the extension by an amount δ. With ATP, the supercoil is removed and the DNA length extends by 2δ. The rate ν of loss of supercoils follows a simple *Michaelis–Menten expression*

$$\nu = V_0 \frac{\Delta Wr}{\Delta Wr + k_{1/2}} \tag{2.31}$$

where the speed of the reaction $V_0 = 2.2$ cycles/s and $k_{1/2} = 2$ supercoils. The speed V_0 has been measured at a force of 0.7 pN as a function of ATP, which can be described by a Michaelis–Menten expression

$$V_0 = \frac{V_{max}[ATP]}{K_M + [ATP]} \tag{2.32}$$

with a K_M value of $K_M = 270 \pm 40\,\mu$M, and $V_{max} = 3.6 \pm 0.2$ turnovers/s, data consistent with bulk measurements. The dependence of V_0 on force was also measured, with the turnover rate decreasing, which indicated that the closure of the cleaved DNA segment might be the rate-limiting factor in the reaction.

We leave the problem of topoisomerases with this qualitative picture. But how does the selection process actually work — how does the enzyme find the position in which to cut? Although many quantitative details about topoisomerase action are known, their precise functioning is still unclear. A proposal for a *kinetic proofreading scenario* was made some time ago (Yan, Magnasco and Marko, 1999). In a kinetic proofreading scenario generally two types of processes occur: first, a recognition step by which the molecule becomes, and then a subsequent ATP-consuming step, i.e., the coupling of a reversible and an irreversible step. Figure 2.16 displays a schematic drawing of this process for Topo II action: one distinguishes between the G-segment, which is the strand at which the enzyme attaches. Collision with the T-segment in step 2 leads to a conformational change (step 3) in which the enzyme becomes activated. From here it can relax back to the initial state 1, or, upon collision with a second T-segment in state 4, perform the strand exchange in step 5. At the moment, we do not have the tools to analyze the properties of this scenario in a precise way — they will be presented in Chapter 5.

Single-molecule studies of nucleosomes

In the previous section we saw the most basic experiments with single-molecule pulling assays. These approaches lend themselves also to the study of protein–DNA interactions, as any protein that binds to DNA influences its behavior under the application of force — the action of Topo II is an excellent example. As a consequence, single-molecule methods have led to a veritable "force spectroscopy" of DNA–protein samples. They can thus also be applied to nucleosomes, both single and multiple, so-called nucleosomal arrays. These experiments are highly complex (even when technical difficulties in the reconstitution of the fibers are surmounted) as the nucleosomal arrays can display numerous features due to the intrinsic rearrangements of the nucleosomes.

Qualitatively what has emerged from these experiments is that for low forces, the force-extension curves resemble those for bare DNA, but with different parameters (Figure 2.17). When the pulling forces become strong enough, typically about 10 pN, a single nucleosome starts to be unwrapped, with one full wrap going first. At higher forces, hysteretic step-like transitions appear (Figure 2.18). We illustrate this with recent results from John van Noort and co-workers (Meng, Andresen, van Noort, 2015) where the observed curves have

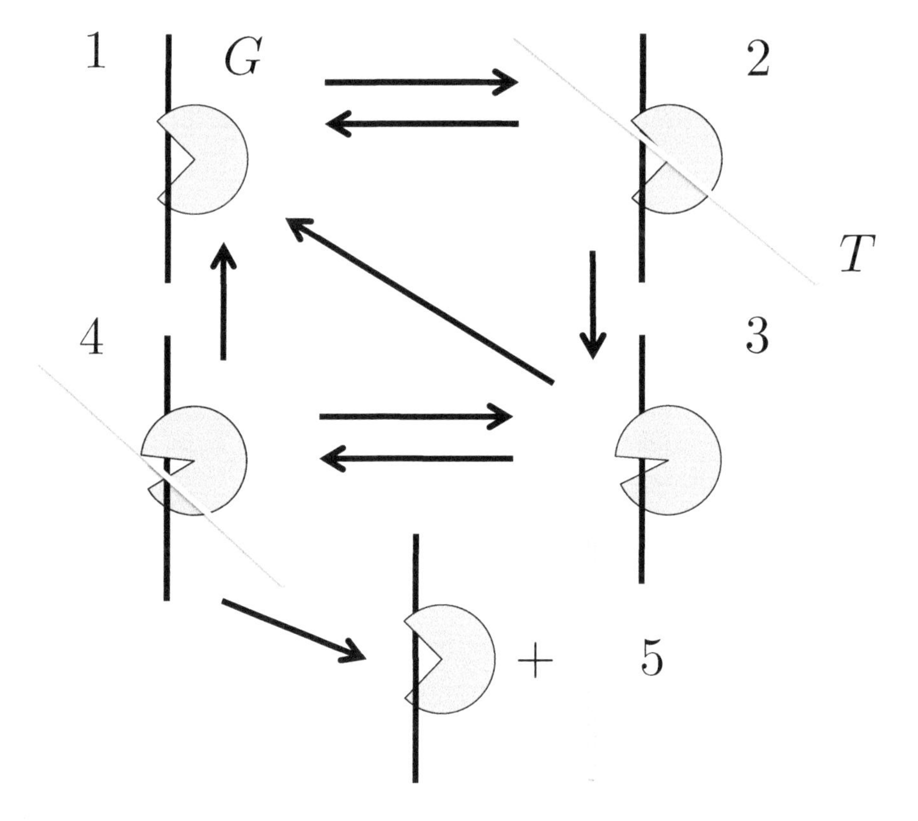

FIGURE 2.16 Kinetic proofreading scenario of topoisomerase strand passage. States (1-5): (1) Topoisomerase II at G-segment; (2) T-segment encounter; (3) activated Topo II; (4) encounter with "correct" T-segment; (5) strand passage. Redrawn after Maxwell et al. (2005) and Bates and Maxwell (2007).

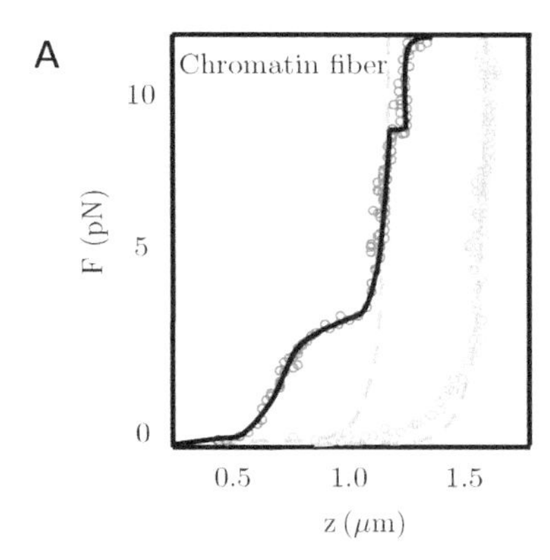

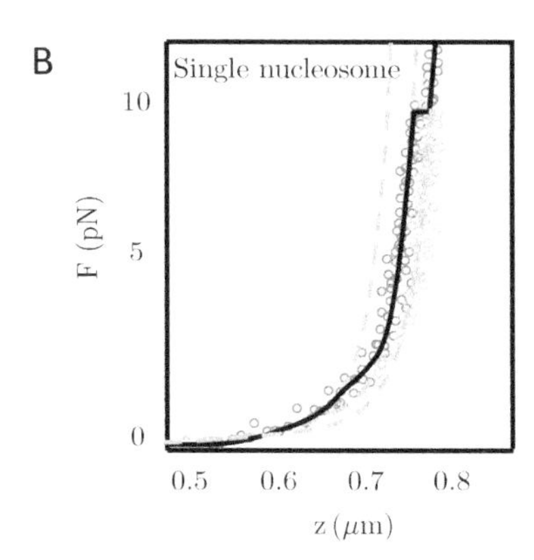

FIGURE 2.17 Force-extension curves of (A) a chromatin fiber and (B) a single nucleosome. Dark: pulling; light grey: release. All curves are reversible until a force of 6 pN is exceeded. The dashed lines refer to the WLC model. The black curves are fitting curves to a model whose parametrization is not detailed here. Reprinted from Meng, Andresen, van Noort (2015), © Oxford University Press.

been interpreted with the aid of an effective fitting model which we do not detail here. Ambiguities of data interpretation remain, as the fairly involved (and controversial) discussion of this problem in the literature shows, but the figures are able to convey a good idea of the kind of intermediate states a chromatin fiber can pass through when it is pulled by a strong force. There is general consensus that the destabilizing forces for the fiber are less than for the DNA itself, and this difference is even greater for the stretching modulus. The chromatin fiber is a much softer object than the underlying DNA itself.

Chromatin fibers beyond 10 nm

DNA extracted from a eukaryotic cell will look like what is shown in the sketch of an electron micrograph of Figure 2.19; for a real EM graph, see, e.g., Boulé, Mozziconacci and Lavelle (2015). We have discussed some of the properties of this 10-nm fiber before. But what does chromatin look like at larger scales? The next step beyond the 10-nm fiber is the 30-nm fiber, which involves the packing of single chromatin fibers. It is, however, an elusive object, as convincing experimental evidence for it could not be found within the cell nucleus. The difficulty in gathering evidence lies in the fact that the 30-nm fiber appears at higher ionic density, and experimental methods that can

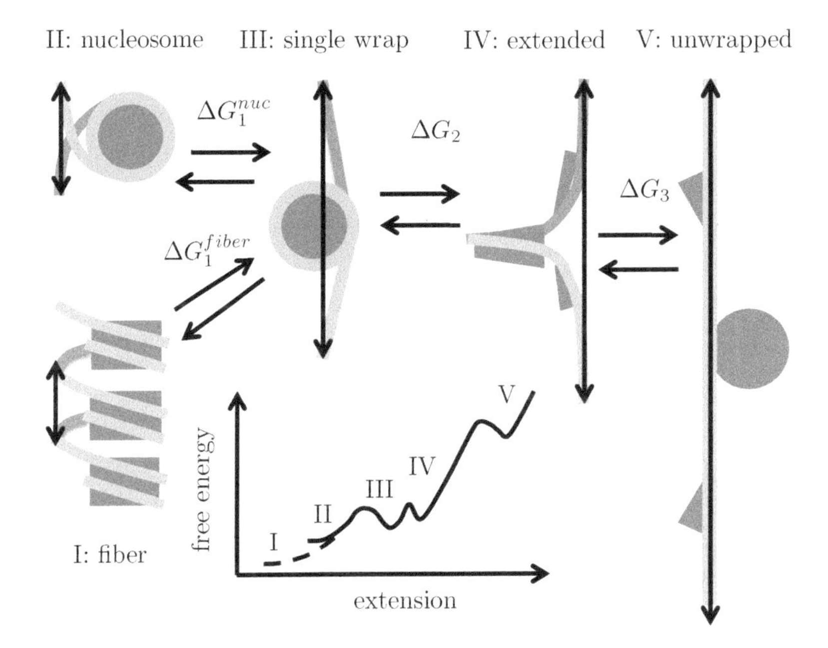

FIGURE 2.18 Schematic drawing of metastable nucleosome conformations encountered upon pulling a chromatin fiber. Upon force increase, first one of the wraps detaches. The dissociation of histones may be associated. Reprinted from Meng, Andresen, van Noort (2015), © Oxford University Press.

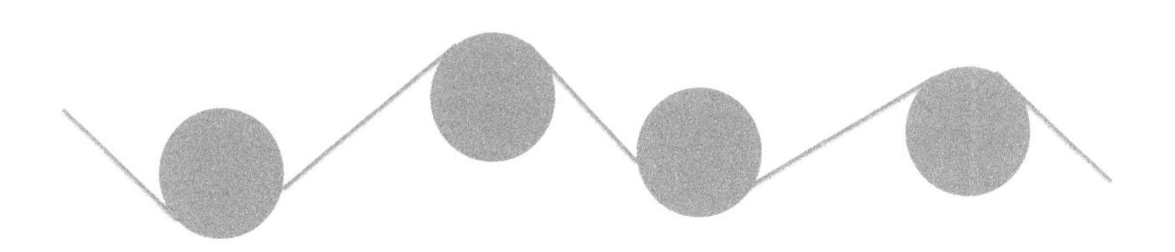

FIGURE 2.19 Sketch of nucleosomes arranged on DNA in an environment of low ionic density: the 10-nm fiber looks like "beads on a string".

address corresponding portions of chromatin within the nucleus with sufficient resolution are currently not available. There are electron micrographs available — see again (Boulé, Mozziconacci and Lavelle, 2015) — but their resolution is not sufficient to unambiguously explain the structure.

This question can in fact more easily be addressed for reconstituted fibers. Song et al. (2014) recently generated such fibers on well-defined DNA templates and characterized their structure with cryo-EM. The observed structure falls in the class of so-called *zig-zag* models of the fiber, whereby the linker DNA moves back and forth between nucleosomes on two superstrands of the chromatin fiber, hence has interactions between sites numbered as $(n, n + 2)$, i.e., from nucleosome n to the *second* following nucleosome. An alternative packing is described by the *solenoid* model, in which the nucleosomes are arranged in a linear way around a superhelix, $(n, n + 1)$, and the linker DNA follows a much smoother path. The two packing structures of the nucleosomes are sketched in Figure 2.20.

Cryo-EM. Cryo-electron microscopy (cryo-EM) is a technique based on transmission electron microscopy (TEM) where the sample is studied at cryogenic temperatures (liquid nitrogen). Cryo-EM allows us to image probes that have not been stained or fixed. The resolution of cryo-EM image is steadily improving, reaching down meanwhile to values < 3 Å. Variants of cryo-EM are cryo-electron tomography (CET) or the random canonical tilt (RCT) method where a 3D reconstruction of a sample is created from tilted 2D images.

Here, we will not go into further detail of the discussion of the possibilities of reconstituted fibers and their structural models and rather refer to the literature at this stage, in particular the book by Schiessel (2014), which gives a detailed discussion of this issue. The reader is also asked to consult the references listed at the end of this chapter. The reason for making this choice here is that, very likely, none of the possible reconstituted structures will be realized in pure form within chromatin, as for all of these structures the associated regulatory proteins are assumed to be absent. The interaction of these proteins and DNA will be our main focus in the following. But first we now take the step upwards to chromatin at the level of the whole nucleus.

Chromatin territories

The basic buildup of the nucleus of a eukaryotic cell is depicted in the sketch of Figure 2.21. The nucleus is enclosed by the *nuclear envelope* of which also its pores are indicated. The darkest structure in the nucleus is called the nucleolus; in EM micrographs one can also distinguish between darker regions

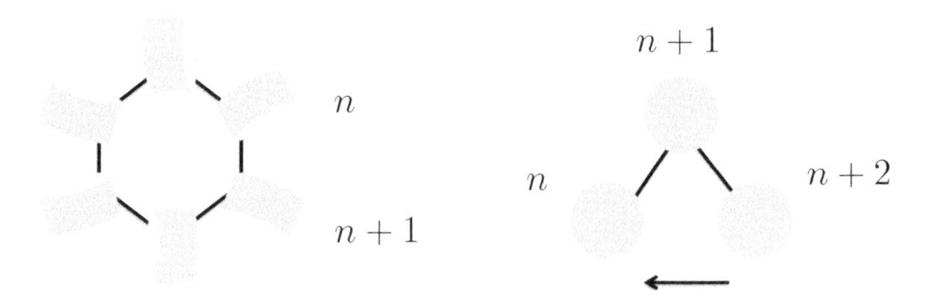

FIGURE 2.20 Left: the packing of the solenoid involves a linear arrangement of nucleosomes ($n, n+1$) around a circle; this circular arrangement is then stacked on top to form the condensed fiber. In the zig-zag packing, shown at the right, the linear nucleosome fiber is arranged such that a nucleosome of index $n+2$ is packed on top of nucleosome n. This packing is then wound into a complete fiber.

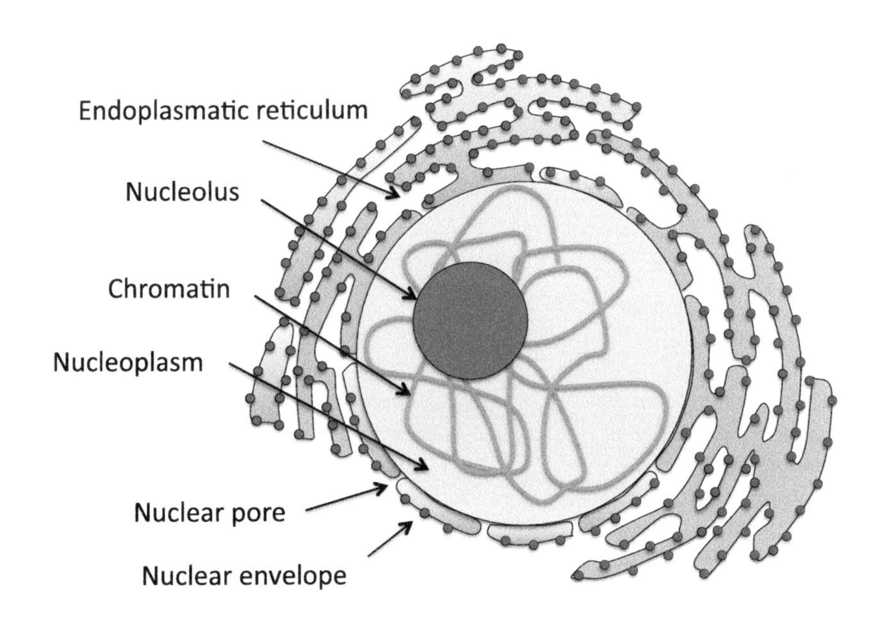

FIGURE 2.21 Schematic buildup of the nucleus of a eukaryotic cell.

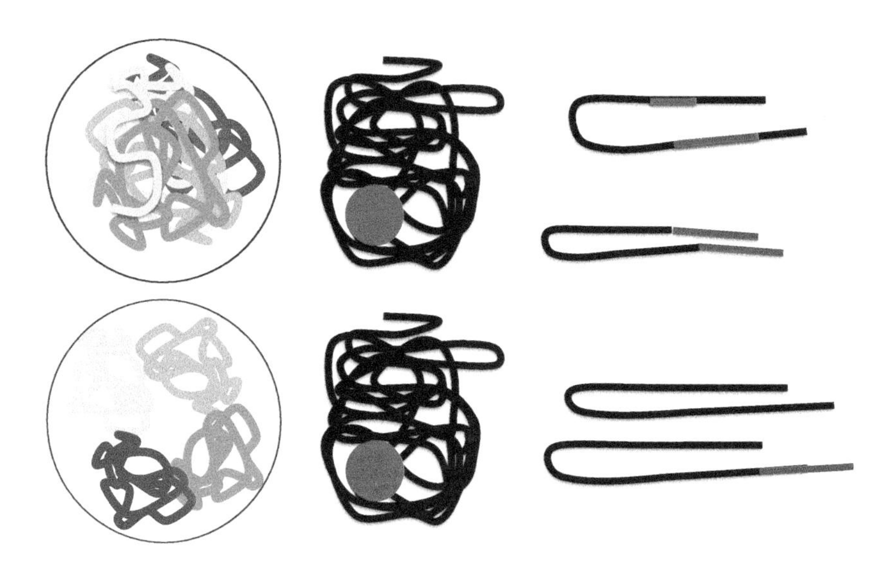

FIGURE 2.22 Rationalization of chromatin territories by a wool model. Left, top: wool threads of different colors intermingle in a spherical space. Left, bottom: wool threads of different colors, mimicking different chromatin territories. Middle: a red spot indicating irradiation of nuclear DNA shown white — the distinction between the mingled and territorial chromatin is not possible. Right: DNA threaded out for both cases. The difference in the distribution of damaged DNA portions is clearly visible. Following Cremer and Cremer (2010).

inside chromatin, the *heterochromatin* and lighter regions, the *euchromatin*, a distinction which is not made in our schematic picture. The *nucleolus* is a multi-functional nuclear compartment in which, among other processes, ribosome synthesis occurs. Finally, the nucleus is surrounded by the endoplasmatic reticulum, which is the site of protein synthesis. The chromatin structure itself, aside from the (still fairly loose) distinction between euchromatin and heterochromatin, is hardly discernible in EM micrographs. It depends on the cell cycle, and decidedly becomes visible when the DNA is packed into the high-density form of condensed chromosomes. But what does it look like, say, in *interphase*, i.e., in the "normal state" of a living cell? The DNA filaments that are the basic building block of chromatin can reach up to 10^8 bp in human cells, which makes them about 30 mm long.

Two hypotheses have dominated the discussion: either all DNA from the different chromosomes is intermingled like spaghetti in a dish, or the interphase chromosomes generally stay always apart in their compartments, or in its technical term, in their *chromatin territory*, CT.

That the latter case indeed prevails can be demonstrated experimentally. It can be nicely illustrated by a simple demonstration, which is shown in Figure 2.22. It is based on wool of different colors put into a spherical confinement (physicist jargon for a bowl or dish). The demonstration is a nice way to grasp the concept, but it also helps to devise a real experiment.

FISH and 3C. Fluorescence in situ hybridization (FISH) is a technique that uses fluorescent probes that bind to chromatin with a high degree of sequence complementarity. Chromosome conformation capture techniques (3C or beyond: 4C, 5C or Hi-C) are techniques employed to analyze the organization of chromatin in the nucleus. They allow us to quantify the number of interactions between genomic loci that are nearby in 3-D space, but may be separated by many nucleotides in the linear genome. The chief difference between 3C-based methods is their scope; in 3C, only the interactions between two specific fragments are quantified. Hi-C methods quantify interaction events between all possible pairs of fragments simultaneously. 3C methods begin with similar steps: first, nuclear DNA is cross-linked, introducing bonds that freeze interactions between genomic loci. The genome is then cut into fragments. Next, random ligation is performed. This quantifies the proximity of fragments, because fragments are more likely to be ligated to nearby fragments.

In the first picture in the top row of images, the wool threads are intermingled — this is, say, the spaghetti phase of chromatin. In the bottom row, the different wool threads remain separate in the first picture. The second picture in both rows has the different colors removed: this is the way the chromatin state appears to the biologist in an experiment. Suppose now we put a red spot of paint arbitrarily on the wool. Where will it be on the wool threads? This is resolved, for the two cases, in the last pictures of the two rows. For the wool threads we see that, in the case of intermingled threads, many of them will carry the paint in some places along the individual threads. In the other case, only a few of the threads will be colored.

Putting such a colored spot means in a real experiment irradiating nuclear DNA with some damaging radiation, damage that can later be made visible. Irradiating the nuclear DNA in a localized spot, the resulting damage would be more or less equally distributed across the chromosomes if they were perfectly intermingled. In reality, the damage is very much localized on just a few chromosomes — a strong argument in favor of chromosome territories,

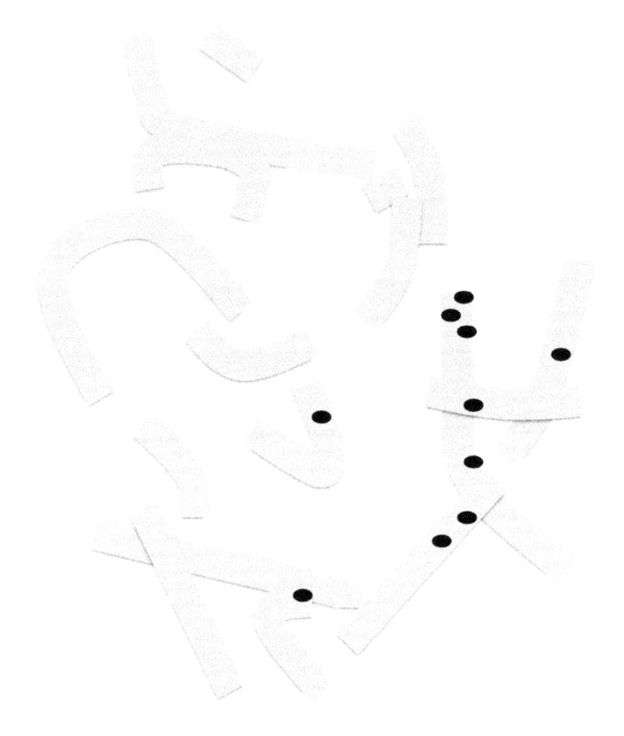

FIGURE 2.23 Sketch of the result of a micro-irradiation experiment on DNA from the nucleus of a Chinese hamster cell. The localization of DNA damage on chromosomes provides direct evidence for the chromosome territory model. Following Cremer and Cremer (2010).

in which each of the chromosomes remains localized. The result of such an irradiation experiment is sketched in Figure 2.23.

Earlier in this chapter we used a simple polymer model, the worm-like chain model (WLC), to discuss the properties of both single DNA molecules as well as of reconstituted fibers. Can we use such models also on the largest scales, for the whole of chromatin in the nucleus? Modern experimental techniques are now available to map out both chromatin structure by fluorescent markers (FISH: fluorescent in situ hybridization) and specialized techniques to mark chromatin contacts (chromosome conformation capture).

We have already seen that the WLC model shows that the assumed linear polymer fiber is stiff on small scales and a random coil bent by thermal fluctuations on large scales. The characteristic length describing the crossover between the two behaviors is related to the persistence length we determined before, and is given typically in terms of the *Kuhn length* ℓ_K, which, for the

WLC model, is twice the persistence length. Its value is $\approx$ 300 nm for the 30-nm fiber. Very generally, the Kuhn length is a measure of the stiffness of a polymer, $\ell_K = L/N$, where L is the *contour length* of the polymer and N is the number of stiff chain elements.

More precisely, let us consider two points on the fiber placed at positions N_1 and N_2 Mbp (Mega base pairs = 10^6 bp) from one end of the fiber. Their respective distance is given by

$$L \equiv |N_2 - N_1| \times 10\,\mu\text{m/Mbp} \tag{2.33}$$

along the fiber contour. One then has the behavior for the *mean-squared spatial distance* in space (i.e., not following the fiber contour, but directly across the three-dimensional volume occupied by the fibers)

$$R^2(L, \ell_K) = \frac{\ell_K^2}{2}\left(\frac{2L}{\ell_K} + e^{-2L/\ell_K} - 1\right). \tag{2.34}$$

For $L \ll \ell_K$, this formula reduces to

$$R^2 = L^2\,, \tag{2.35}$$

meaning that the fiber is essentially stiff, while for $L \gg \ell_K$, the fiber is easily bent by thermal effects, and one has

$$R^2(L) = \ell_K L\,. \tag{2.36}$$

Figure 2.24 displays the mean-squared distance for chromatin of different organisms (yeast, drosophila, human) (A. Rosa and R. Everaers, 2008; A. Rosa, 2013). The value at lowest L corresponds to yeast and falls onto this curve. For larger L, hence the larger genomes of higher organisms, the data deviate and are rather better described by a behavior with a different power law,

$$R^2 \sim L^{2/3}\,. \tag{2.37}$$

Such behavior can be captured by other types of polymer models, either assuming that the chromatin structure corresponds to "loops" on a chain, with a presumed length of such loops at about 10^5 bp, or invoking ring-like polymer structures. The bottom graph in Figure 2.24 shows the contact frequencies of chromatin fibers for yeast and human for which also an analytic expression exists for the WLC model. For large L, this contact frequency decays as $L^{-3/2}$, whereas the more complex models can reproduce the observed L^{-1} decay for human chromatin (A. Rosa, 2013).

The lesson to be learned from this result is that the large-scale behavior of chromatin, and also its partitioning chromatin territories, does fundamentally reflect the properties of DNA (or the chromatin fiber) as being a polymer, and

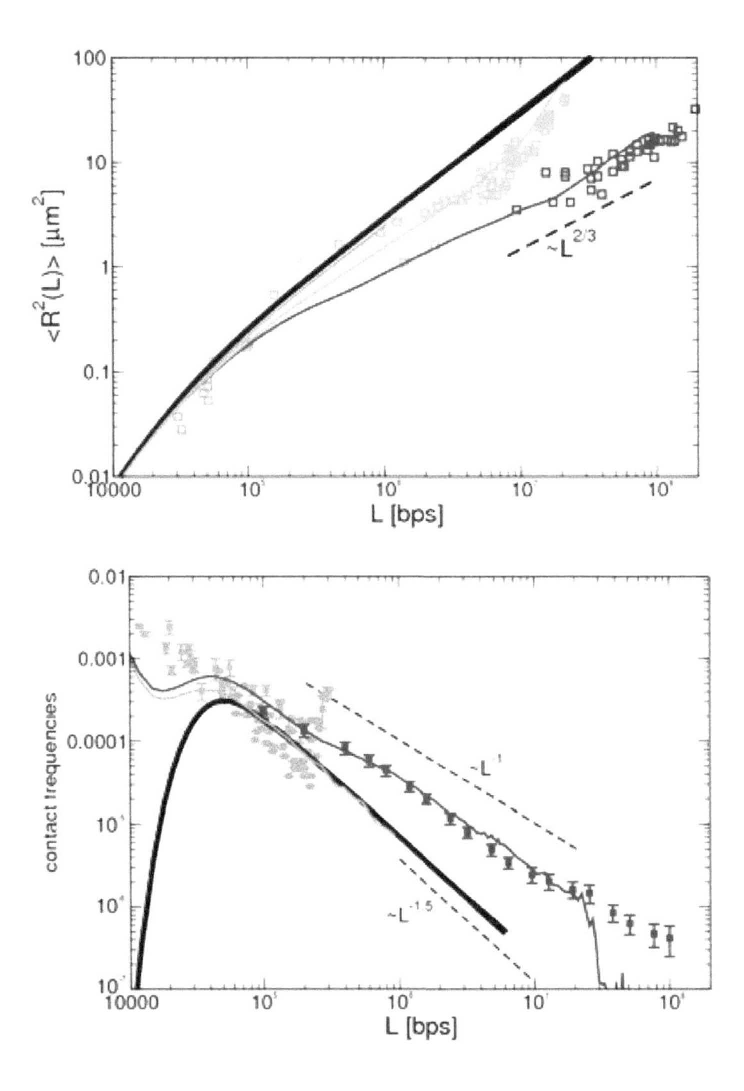

FIGURE 2.24 Top: mean-squared distances as a function of length L. Bottom: Contact frequencies between chromatin fibers. Reprinted with permission from Rosa (2013). © Portland Press.

TABLE 2.1 Characteristic mechanical data of DNA and the chromatin fiber. Reprinted with permission from C. Lavelle (2014), © Elsevier. For the original sources from which these data were taken, please consult this paper.

Property	Data
DNA length stretched/random coil radius	2 m; 130 mm
DNA compaction (volume reduc.; packing fr.)	2×10^4
Chromatin length (30-nm fiber)	50 mm
DNA bending persistence length	50 nm ($\sim$ 150 bp)
DNA torsional persistence length	40–100 nm
DNA stretching modulus	1100 pN
ssDNA bending persistence length	0.7 nm
Force required for B to S DNA transition	65 pN
Force required for DNA breakage	500 pN
Force for unfolding of hairpins and 2d structures	13 pN
Chromatin bending persistence length	30–200 nm
Chromatin torsional persistence length	5 nm
Chromatin stretching modulus	5–8 pN
Force required to disrupt a nucleosome	2–20 pN
Torque required to disrupt a nucleosome	36 pN nm
RNA polymerase stalling torque	15–25 pN
Torque required for DNA denaturation	8–10 pN
Force required for chromatin fiber unfolding	4–5 pN
Chromatin resistive torque (plectonemic regime)	3–7 pN
DNA resistive torque (plectonemic regime)	5–30 pN
RNA polymerase transcription rate	10 b/s–100 b/s
Topoisomerase I relaxation kinetics	$\leq$ 100 supercoils/s
Topoisomerase II relaxation kinetics	3-6 supercoils/s

it does not require specific intermolecular interactions between the different chromosomes. Clearly, there is regulatory crosstalk between genes located on different chromosomes, but the basic underlying structure of chromatin within the nucleus can be well captured by basic models from the statistical physics of polymers.

We close this chapter with a table collecting a number of experimentally measured quantities on the action of forces at DNA, the nucleosome, and the chromatin fiber, Table 2.1.

References

Bates A.D. and A. Maxwell
Energy coupling in type II topoisomerases: why do they hydrolyze ATP?
Biochemistry 46, 7929-7941 (2007)

Boulé J.-B., J. Mozziconacci and C. Lavelle
The polymorphisms of the chromatin fiber.
J. Phys.: Condens. Matter 27, 033101 (2015)

Bustamante C., S.B. Smith, J. Liphardt and D. Smith
Single-molecule studies of DNA mechanics.
Curr. Op. Struct. Biol. 10, 279-285 (2000)

Cremer T. and M. Cremer
Chromosome Territories.
Cold Spring Harbour Perspectives in Biology, doi: 10.1101/cshperspect.a003889 (2010)

Lavelle C.
Pack, unpack, bend, twist, pull, push: the physical side of gene expression.
Curr. Op. Gen. & Dev. 25, 74-84 (2014)

Maxwell A., L. Costenaro, S. Mitelheiser and A.D. Bates
Coupling ATP hydrolysis to DNA strand passage in type IIA DNA topoisomerases.
Biochem. Soc. Transact. 33, 1460-1464 (2005)

Meng H., K. Andresen and J. van Noort
Quantitative analysis of single-molecule force spectroscopy on forced chromatin fibers.
Nucl. Acids. Res. 43, 3578-3590 (2015)

Mirkin S.M.
DNA topology: fundamentals
Encyclopedia of Life Sciences, 1-11 (2001)

Rosa A.
Topological constraints and chromosome organization in eukaryotes: a physical point of view.

Biochem. Soc. Transact. 41, 612-615 (2013)

Rosa A. and R. Everaers
Structure and dynamics of interphase chromosomes.
PLoS Comput. Biol. 4, e1000153 (2008)

Schiessel H.
Biophysics for Beginners. A journey through the cell nucleus.
Pan Stanford Publishing (2014)

Song F., P. Chen, D. Sun, M. Wang, L. Dong, D. Liang, R.-M. Xu, P. Zhu and G. Li
Cryo-EM study of the chromatin fiber reveals a double helix twisted by tetranucleosomal units.
Science 344, 376-380 (2014)

Strick T.R., J.-F. Allemand, D. Bensimon, A. Bensimon and V. Croquette
The elasticity of a single supercoiled DNA molecule.
Science 271, 1835-1837 (1996)

Strick T.R., V. Croquette and D. Bensimon
Single-molecule analysis of DNA uncoiling by a type II topoisomerase.
Nature 404, 901-904 (2000)

Yan J., M.O. Magnasco and J.F. Marko
A kinetic proofreading mechanism for disentanglement of DNA by topoisomerases.
Nature 401, 932-935 (1999)

Further reading

Brock Fuller F.
The writhing number of a space curve.
Proc. Natl. Acad. Sci. (USA) 68, 815-819 (1971)

Crick F.H.C.
Linking numbers and nucleosomes.
Proc. Natl. Acad. Sci. (USA) 73, 2639-2643 (1976)

These two papers are classics in the field, worth reading for general education.

Bancaud A., N. Conde e Silva, M. Barbi, G. Wagner, J.-F. Allemand, J. Mozziconacci, C. Lavelle, V. Croquette, J.-M. Victor, A. Prunell and J.-L. Viovy
Structural plasticity of single chromatin fibers revealed by torsional manipulation.
Nat. Struct. Mol. Biol. 13, 444-450 (2006)

A highly detailed experimental and theoretical study of the mechanics of chromatin fibers, following earlier work by

Brower-Toland B.D., C.L. Smith, R.C. Yeh, J.T. Lis, C.L. Peterson and M.D. Wang
Mechanical disruption of individual nucleosomes reveals a reversible multi-stage release of DNA.
Proc. Natl. Acad. Sci. 99, 1960-1965 (2002)

and

Cui Y. and C. Bustamante
Pulling a single chromatin fiber reveals the forces that maintain its higher-order structure.
Proc. Natl. Acad. Sci. 97, 127-132 (2000)

Bustamante C., Z. Bryant and S.B. Smith
Ten years of tension: single-molecule DNA mechanics.
Nature 421, 423-427 (2003)

A still highly readable account of single-molecule techniques by a pioneer in the field.

Lavelle C., J.-M. Victor and J. Zlatanova
Chromatin fiber dynamics under tension and torsion.
Int. J. Mol. Sci. 11, 1557-1559 (2010)

Marko J.F.
Biophysics of protein-DNA interactions and chromosome organization.
Physica A 418, 126-153 (2015)

A concise, quantitative review of the biophysical properties of DNA and pro-

teins.

Martínez-García B., X. Fernández, O. Díaz-Ingelmo, A. Rodríguez-Campos, C. Manichanh and J. Roca
Topoisomerase II minimizes DNA entanglements by proofreading DNA topology after DNA strand passage.
Nucl. Acids Res. 42, 1821-1830 (2014)

The following papers contain further information on chromatin structure:

Dekker J., K. Rippe, M. Decker and N. Kleckner
Capturing chromosome conformation.
Science 295, 1306-1311 (2002)

Emanuel M., N.H. Radja, A. Henriksson and H. Schiessel
The physics behind the large-scale organization of DNA in eukaryotes.
Phys. Biol. 6, 025008 (2009)

Korolev N., Y. Fan, A.P. Lyubartsev and L. Nordenskiöld
Modelling chromatin structure and dynamics: status and prospects.
Curr. Op. Struct. Biol. 22, 151-159 (2012)

O'Sullivan J.M., M.D. Hendy, T. Pichugina, G.C. Wake and J. Langowski
The statistical mechanics of chromosome conformation capture.
Nucleus 4, 390-398 (2013)

Ozer, G., A. Luque and T. Schlick
The chromatin fiber: multiscale problems and approaches.
Curr. Op. Struct. Biol. 31,124-139 (2015)

Pederson T., M.C. King and J.F. Marko
Forces, fluctuations, and self-organization in the nucleus.
Mol. Bio. Cell 26, 3915-3919 (2015)

Woodcock C.L. and R.P. Ghosh
Chromatin Higher-order Structure and Dynamics.
Cold Spring Harbour Perspectives in Biology, doi:10.1101/cshperspect.a000596 (2010)

CHAPTER 3

Gene regulation without and with chromatin

Gene regulation: lessons from a phage

In order to understand how genes are regulated in chromatin we first need to understand how they are regulated *without*. Gene regulation without chromatin essentially means gene regulation in bacteria which lacks the nucleosomal level of compaction and regulation. For our purpose here, we can assume that DNA in this case is freely accessible for regulatory molecules, although this is not entirely correct, as there are mechanisms that restrict access of regulatory molecules to DNA also in bacteria, e.g., supercoiling or active compaction by proteins.

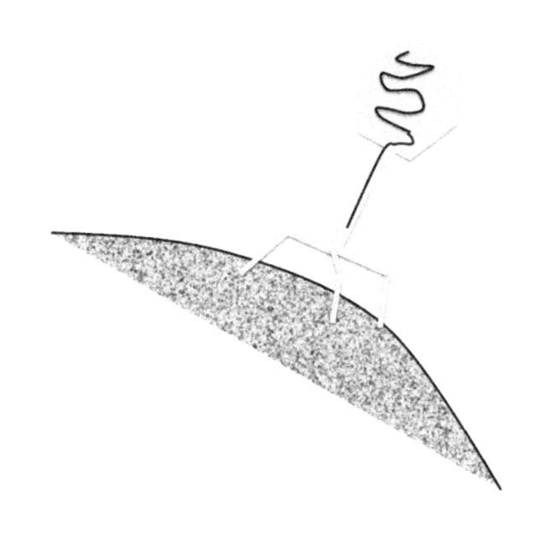

FIGURE 3.1 A λ-phage infecting an *E. coli* bacterium.

The mechanism of gene regulation in eukaryotes involves numerous steps. While the central dogma of molecular biology as formulated by (or ascribed to) Francis Crick, "DNA makes RNA makes protein", is also the basis of the gene regulation in eukaryotes, but several additional mechanisms need to be taken into account. Here we are first interested in the initial steps of gene expression, which means, in a loose sense, "making the gene ready for the transcription by polymerase".

Instead of looking at bacteria, we describe the regulation of genes in an even simpler organism, which is a phage: the famous λ-phage which infects *E. coli* bacteria. Figure 3.1 shows the infection of an *E. coli* bacterium by λ-phages, which punch a hole into the bacterial membrane and insert a DNA plasmid, a circular DNA molecule, into the cell (Ptashne, 2004).

The infection brings the λ-phage genome into the cell where it can capture the cell's machinery for transcription. The λ-phage genome has a built-in switch between two lifestyles whose states determine the read-out of two neighboring genes. The regulatory element concerned with this decision is called the right operator, O_R. It consists of three DNA sites which are recognized by two regulatory proteins, called the λ-repressor and the protein cro. When the phage is in the "lysogenic state", in which the infection stays dormant, the product of the gene *cI* is synthesized, which is the λ-repressor, and it occupies the sites O_R1 and O_R2. In this configuration, the transcription of the *cro* gene is suppressed.

If the λ-repressor molecule is destroyed through cleavage by the RecA protein, a process induced by DNA damage, the *cro* gene becomes unblocked and is transcribed. The resulting protein cro then blocks site O_R3, switching off the transcription of the *cI* gene.

Figure 3.2 shows a part of this process: the interaction of the λ-repressor with the operator sites. One important step is the formation of dimers of the repressor molecules, another the interaction of bound dimers with neighboring dimers. Both processes are examples of mechanisms showing cooperativity, i.e., the increase of the strength of an interaction by complex formation.

Table 3.1 displays the energetics of regulatory control at the operator module between the two genes *cI* and *cro*. Note that for the configurations in which repressor dimers R occupy neighboring sites, the free energy ΔG is augmented by a contribution due to cooperativity between the dimers. Based on these free energies, the probability of an operator being in the s configuration is given by

$$f_s = \frac{\exp(-\Delta G_s/RT)[R_2]^j}{\sum_{s,j}\exp(-\Delta G_s/RT)[R_2]^j} \tag{3.1}$$

where $[R_2]$ is the concentration of unbound repressor dimers, and $j(s)$ the number of repressor dimers bound to an operator in configuration s. The total concentration of repressors is

$$[R_t] = [R_1] + 2[R_2] + 2[O_t]\sum j f_s \tag{3.2}$$

where $[R_1]$ is the concentration of repressor monomers which cannot bind the operator sites, and $[O_t]$ is the total concentration of operators. The binding of repressor monomers to dimers is given by $[R_2] = K_a[R_1]^2$.

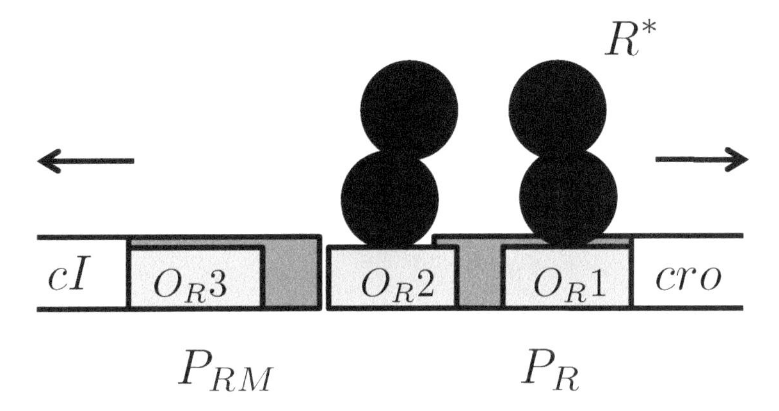

FIGURE 3.2 Regulatory programming of an operator segment in λ-phage, following Ackers, Johnson and Shea (1982). Note the structure of the regulatory unit: two promoters adjacent to two genes that are transcribed in opposite directions, with three partially overlapping operator sites. (Promoters in dark grey, operators in light grey.) O_R3 overlaps with promoter P_{RM}, O_R2 (partially) and O_R1 with P_R. Shown is configuration $s = 7$ from Table 3.1, in which two repressor dimers R bind to the operator sites, and also mutually reinforce their interaction, hence the label *.

TABLE 3.1 Energetics of the regulatory module. s: Species, O: vacant site, R: site occupied by repressor dimer, R*: dimer in cooperative interaction. The free energy (FE) per binding site is given by $\Delta G_s = -RT \ln K_s$ for $s = 1, 2, 3$ where K_s is the equilibrium constant of dimer binding and R the gas constant. 1 kcal = 4.18 J. Data from Ackers, Johnson and Shea (1982).

s	O_R1	O_R2	O_R3	Free energy	Total FE in kcal
1	O	O	O	Reference state	0
2	O	O	R	ΔG_1	-11.7
3	O	R	O	ΔG_2	-10.1
4	R	O	O	ΔG_3	-10.1
5	O	R*	R*	$\Delta G_1 + \Delta G_2 + \Delta G_{12}$	-23.8
6	R	O	R	$\Delta G_1 + \Delta G_3$	-21.8
7	R*	R*	O	$\Delta G_2 + \Delta G_3 + \Delta G_{23}$	-22.2
8	R	R*	R*	$\Delta G_1 + \Delta G_2 + \Delta G_3 + \Delta G_{12}$	-33.9

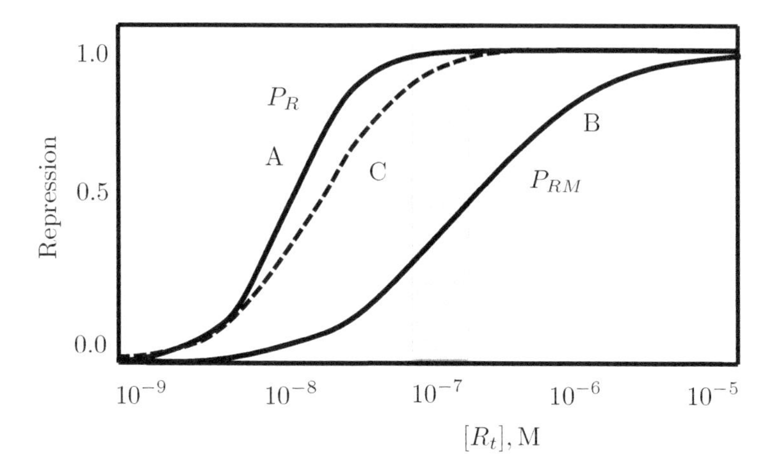

FIGURE 3.3 Predicted repressing curves $f_{P_{RM}}$ and f_{P_R} as a function of total repressor concentration $[R_t]$. For discussion, see text. Redrawn from (Ackers, Johnson and Shea, 1982).

The binding to the operator sites controls the effect of the promoters on the two genes, P_R of *cro* and P_{RM} of *cI*. P_{RM} is controlled by O_R3 only, and hence determined by

$$f_{O_R3} = f_{P_{RM}} = f_4 + f_5 + f_7 + f_8 \tag{3.3}$$

while P_R is controlled by both O_R1 and O_R2:

$$f_{O_R3} = f_{P_{RM}} = f_2 + f_3 + f_5 + f_6 + f_7 + f_8\,. \tag{3.4}$$

The predicted repression curves are shown in Figure 3.3 as a function of total repressor concentration, in which the curves A and B represent Eqs. (3.4) and (3.3), respectively. Curve C corresponds to Eq. (3.4), neglecting the cooperative effect, which shows indeed that its effect is to sharpen the response. The shaded region corresponds to the regime of repressor concentration found in a lysogenic cell.

This discussion shows that the control of promoters is due to a combination of three effects: (i) the build-up of the promoter-operator region which determines the combinatorial possibilities (i.e., the regulatory wiring); (ii) the binding energy of the dimers to the operator regions; (iii) cooperative interactions both in the binding of the repressor monomers to dimers, and then the cooperative effect of binding dimers on neighboring operator sites.

These mechanisms are the basis for all gene regulatory processes in bacteria.

Given the presence of nucleosomes in chromatin, it cannot be sufficient in eukaryote organisms. We will thus learn about the mechanisms that Nature has come up with to tackle the nucleosomes: the first step is to understand where the nucleosomes go depending on the underlying DNA sequence in relation to the encoded transcriptional information, taking into account our previously obtained knowledge about where a single nucleosome would prefer to go, and how this is influenced by the other nucleosomes.

Genome-scale determination of nucleosome positions

From the previous section we can draw the conclusion that specific sequences upstream of a gene exert control over the downstream gene, and the mutual blocking of the sites by transcription factors is an important regulatory mechanism. In some sense nucleosomes — although structurally much more complex — can also serve as "transcription factors" — either by being in the way of other regulators, or giving access to regulatory sequences and the gene by being absent from such sequences. In order to pass from prokaryote to eukaryote it is thus important to know where the nucleosomes are. We know of two factors already: one is their intrinsic sequence preference based on DNA bending, and the second is a purely statistical effect due to their mutual avoidance. Are these two effects enough to predict where they are on a DNA sequence of genomic length? Probably not, as this entirely ignores the encoded biological meaning of the sequences.

Nucleosomes do avoid poly(dA:dT) tracts along the DNA sequence, as these are among the least bendable sequence elements. The natural variation of these tracts along the genome therefore does give strong hints at where nucleosomes will *not* be located. On the other hand, strongly positioning sequences are found in the vicinity of promoters, and in exon-rich regions. However, in vivo, numerous additional factors play a role, as the chromatin fiber is not a static object and under permanent activity of numerous molecules, mostly proteins. Therefore a strategy has been adopted to generate an in vitro map of nucleosome positions on the yeast genome, and to compare this result with in vivo data. Several predictive software programs have been developed for this purpose, and despite several different input assumptions yield fairly comparable results (Liu et al., 2014).

We here only describe the simplest of these approaches, as it depends on the least number of parameters (Van der Heijden et al., 2012). The model is built with the intention to test the hypothesis that dinucleotide probability functions are sufficient to predict the nucleosome binding preferences — their free energy landscape. This computation therefore appears in a similar vein as the model by Ackers, Johnson and Shea for the binding of transcription factors — but on a genomic scale.

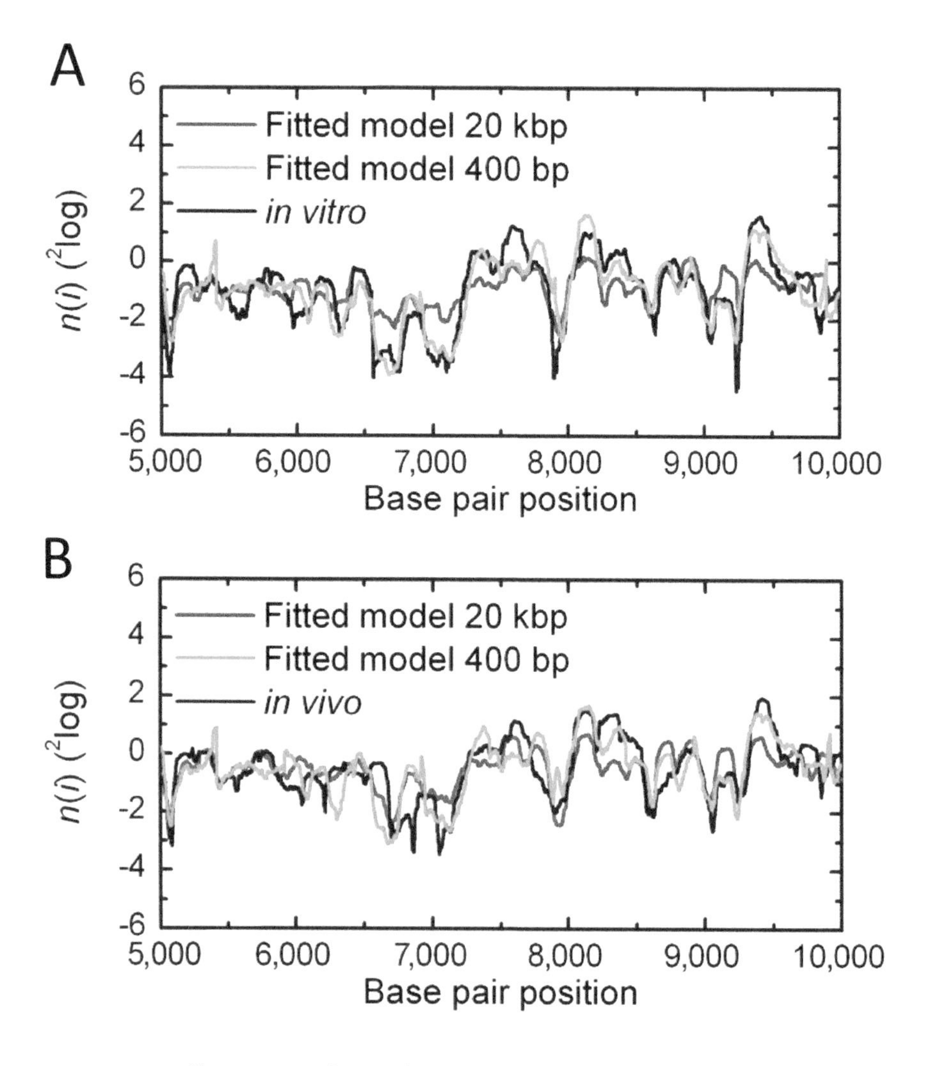

FIGURE 3.4 Genome-wide nucleosome occupancy for in vitro (A) and in vivo (B) data compared to the present model fitted with two windows. Reprinted with permission from Van der Heijden et al. (2012). © PNAS.

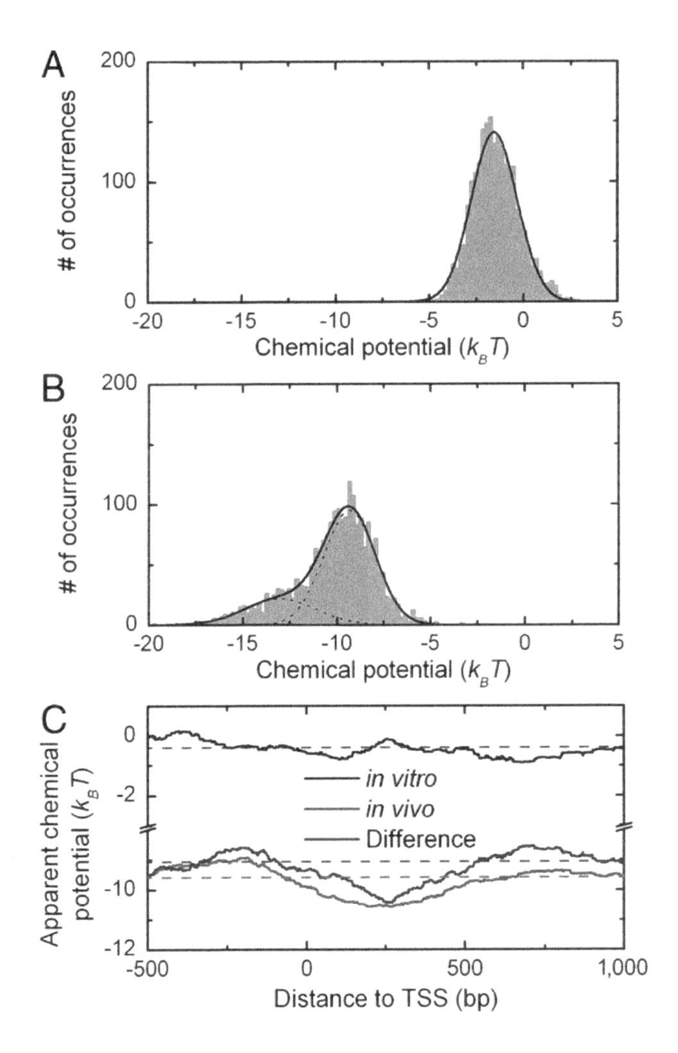

FIGURE 3.5 Distributions of the apparent chemical potential for the in vitro (A) and in vivo (B) data. While the top distribution is a normal distribution with a standard deviation of 1 k_BT, the bottom distribution displays a distinct shoulder, identified with nucleosomes of higher affinity. (C) The average apparent chemical potential around transcription start sites (TSS) in the yeast genome to distinguish the differences between in vitro and in vivo data. For example, the difference (blue) in the apparent chemical potential between shows a clear decrease at the start of the open reading frame, indicating that the strongly positioned nucleosome +1 is not solely defined by the DNA sequence rules of the model. Reprinted with permission from Van der Heijden et al. (2012). © PNAS.

The authors model the periodic distribution of the dinucleotides AA, TA, TT, which mainly affect DNA bending, with a 10-bp periodicity, and of GC, which mainly affects twist, with an additional relative 5-bp shift according to the conditional probability for the nucleotide Y at position s relative to the dyad axis

$$P(Y_s|Y_{s-1}) = P_0 + B\cos\left(2\pi\left(\frac{s}{p} + \frac{1}{2}\delta_{GC}\right)\right) \tag{3.5}$$

for nucleotide $Y_s = \{A, C, G, T\}$. $P_0 = 0.25$ is the basal probability of finding one of the nucleotides. Here, p is periodicity, and the preferred orientation of GC outward of the nucleosome, and of the other dinucleotides to face inward is taken into account by the Kronecker-δ function, which is 1 for GC and 0 otherwise. The conditional probability is normalized over Y_s,

$$\sum_{Y_s} P(Y_s|Y_{s-1}) = 1\,. \tag{3.6}$$

The likelihood ratio $P_{nuc}(i, N)$ to find a nucleosome dyad at base pair i in a sequence of length N is then given by

$$P_{nuc}(i, N) = 4^N \prod_{s=i-N/2}^{i+N/2} P(Y_s|Y_{s-1})\,. \tag{3.7}$$

The parametrization requires fixing the three parameters periodicity p, amplitude B and sequence length N, which is optimized for $N = 74$, corresponding to 147 bp, $B = 0.2$ and $p = 10.0$ bp.

The likelihood results can be converted to a free energy landscape via the expression

$$\Delta G_i = -k_B T \ln P_{nuc}(i, N)\,. \tag{3.8}$$

An important fine point is that for the computation a weighted average over both strands needs to be considered, as the conditional probability is not symmetrical for the dinucleotides AT and CG. Further, the final result is passed through a 10-bp low pass filter.

These sequence-dependent data for ΔG are then used to calculate the equilibrium density of histone octamer positions $n(i)$ via the *Percus equation of state* (Vanderlick, Scriven and Davis, 1986),

$$\beta\mu = \beta\Delta G(i) + \ln n(i) - \ln N(i, \sigma) + \int_{i-\sigma}^{i} dx' \frac{n(x')}{N(x', \sigma)} \tag{3.9}$$

where

$$N(a, \sigma) = 1 - \int_a^{a+\sigma} dx' n(x')\,. \tag{3.10}$$

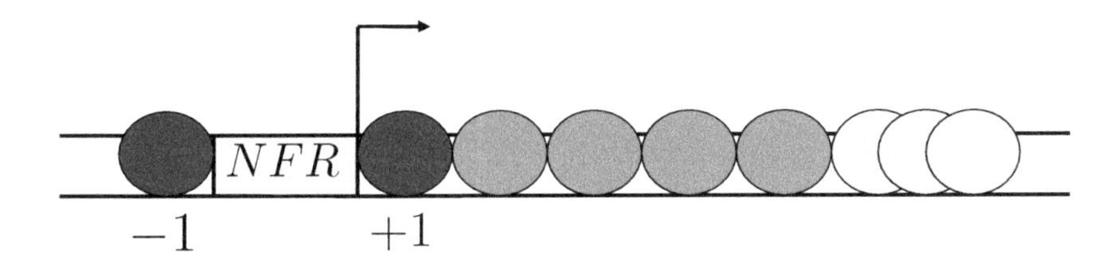

FIGURE 3.6 Nucleosome positioning at a housekeeping gene. Drawn after Radmann-Livaja and Rando (2010).

Here μ is the chemical potential of the histone–DNA interaction and σ the footprint of a histone octamer, both additional fitting parameters.

The results for the nucleosome occupancy are shown in Figure 3.4 for in vitro and in vivo data. Two different fitting windows are indicated. Figure 3.5 shows the chemical potential for the two datasets, which yields a Gaussian distribution for the in vitro data, while the in vivo data exhibit a shoulder, representing nucleosomes with a higher affinity. The comparison of chemical potentials across the genome relative to transcription start site positions (TSS) shows a decrease at the start of the open reading frame (ORF, see box) which can be seen as an indication that the strongly positioned nucleosome +1 is not purely determined by the assumed dinucleotide probabilities.

Open reading frame (ORF). An open reading frame is a continuous stretch of codons that do not contain a stop codon (usually UAA, UAG or UGA) and hence has the potential to code for a protein.

The overall picture that emerges from these studies can be summarized in the following simplified drawing of the positioning of nucleosomes at a general "housekeeping" gene, Figure 3.6. With the notion of "housekeeping gene" one refers to a gene that is constitutively accessible, as its expression is regularly needed in the course of the cell cycle.

Back to the housekeeping gene, we see in Figure 3.6 that two nucleosomes are strongly positioned by sequences that favor nucleosomes, at positions −1 and +1 around a nucleosome free region (NFR) which essentially coincides with the promoter region of the downstream gene. The nucleosomes downstream of this gene are positioned due to the statistical positioning effect, and this order dies out when progressing into the gene sequence. However, the results from the genome positioning study show that while sequence is relevant, there is more biology needed to determine the position of the strongly positioned nucleosome.

In order to round up this story it is now about time to bring the histone tails into the picture, which we have entirely neglected so far. But before doing so in the next chapter, we make a digression into the dynamics of a gene, within a simplified model, which nevertheless requires some fairly complex mathematics in order to analyze it.

Self-regulation of a single gene — a mathematical model

In this section we take a look at the simplest possible case of the dynamics of a gene: a single gene that regulates its own output. This type of gene is indeed an abstraction of the λ-phage genes, neglecting all intertwined regulation levels with its neighboring gene in the λ-switch we have discussed before. We base our following discussion on the gene gate model (GGM), which has been developed as an abstract description for (mainly) small gene networks (Blossey, Cardelli and Phillips, 2006; 2008). The discussion given here follows (Vandecan and Blossey, 2013).[1]

In our context, the gene gate model comes in two variants describing a repressed and an activated gene, respectively. In the repressed case (R) we assume that the gene G can be constitutively transcribing at a basal rate ε,

$$G \to_{\varepsilon} G + P\,, \tag{3.11}$$

which is considered the "on"-state of the gene. The resulting proteins degrade at a rate δ,

$$P \to_{\delta} 0\,. \tag{3.12}$$

Obviously, we have neglected the translation step; it is not much effort to put it also into the model — but its solution then becomes much more complex.

The feedback of protein P on the gene can affect gene activity by putting the gene into its second state. This is described by the reaction

$$G + P \to_{r} G' + P\,. \tag{3.13}$$

In the repressed case, the gene state G' is not productive and simply relaxes

[1]A warning: this part is for the more mathematically inclined. It tries to make a specific point about the properties of stochastic models in gene regulation and the relation to cooperative effects we discussed before. Readers interested in more immediately biologically relevant questions should feel free to skip over it.

back to G at a rate η,

$$G' \rightarrow_{\eta} G\,. \tag{3.14}$$

G' is thus the "off"-state of the gene. In the activated case (A), the gene is productive in state G' via

$$G' \rightarrow_{\eta} G + P \tag{3.15}$$

hence it relaxes back to gene state G while producing protein P, if the rate $\eta > \varepsilon$. Here, G and G' change roles: G is now the "off"-state (although never fully off) and G' is the "on"-state. We note that in the case $\eta < \epsilon$, the gene is actually repressed, but left with protein production at a finite rate η. This case does, in fact, not differ qualitatively from the case under (R).

This simple set of reactions with variables G, G', P and parameters $(\varepsilon, \delta, r, \eta)$ corresponds in terms of its complexity to a model introduced very early by Peccoud and Ycart (1995), which we abbreviate by PYM. The major difference between the present model and the PYM is that the gene in the PYM never interacts with its proteins and reaction (3.13) occurs without the intervention of proteins. Further, the reaction (3.15) of case (A) does not apply. The main difference between the GGM discussed here and the PYM lies, in fact, precisely in this reaction (3.15), as it leads in the master equation to state changes involving both the gene state and the protein number.

The master equations for the two versions of the GGM are readily written down. Denoting the basal state of the gene by 0, its second state by 1 and counting the protein number by n, one can introduce time-dependent probability distributions $p_{i,n}(t)$, with $i = 0, 1$. In case (R), (3.14) applies and the master equations read as

$$\begin{aligned}
\forall n \geq 0 : \partial_t p_{1,n} &= nrp_{0,n} - \eta p_{1,n} + \delta[(n+1)p_{1,n+1} - np_{1,n}] \\
\partial_t p_{0,0} &= \eta p_{1,0} - \varepsilon p_{0,0} + \delta p_{0,1} \\
\forall n \geq 1 : \partial_t p_{0,n} &= \eta p_{1,n} + \varepsilon p_{0,n-1} - (\varepsilon + nr)p_{0,n} \\
&\quad + \delta[(n+1)p_{0,n+1} - np_{0,n}]
\end{aligned} \tag{3.16}$$

while in case (A), for which Eq. (3.15) applies, we have

$$\begin{aligned}
\partial_t p_{0,0} &= -\varepsilon p_{0,0} + \delta p_{0,1} \\
\forall n \geq 1 : \partial_t p_{0,n} &= -(\varepsilon + nr)p_{0,n} + \varepsilon p_{0,n-1} + \eta p_{1,n-1} \\
&\quad +\delta[(n+1)p_{0,n+1} - np_{0,n}]
\end{aligned}$$

$$\forall n \geq 0 : \partial_t p_{1,n} = -\eta p_{1,n} + nrp_{0,n} + \delta[(n+1)p_{1,n+1} - np_{1,n}] \,. \tag{3.17}$$

Introducing the generating functions

$$G_0(z,t) = \sum_{n=0}^{\infty} p_{0,n}(t) z^n \tag{3.18}$$

and

$$G_1(z,t) = \sum_{n=0}^{\infty} p_{1,n}(t) z^n \,, \tag{3.19}$$

the master equations above are transformed into first-order differential equations in t and z. For case (R), we have

$$\begin{aligned} \partial_t G_0 &= \eta G_1 + \varepsilon(z-1)G_0 - (\delta(z-1) + rz)\partial_z G_0 \\ \partial_t G_1 &= -\eta G_1 - \delta(z-1)\partial_z G_1 + rz\partial_z G_0 \end{aligned} \tag{3.20}$$

while in case (A), we have

$$\begin{aligned} \partial_t G_0 &= \eta z G_1 + \varepsilon(z-1)G_0 - (\delta(z-1) + rz)\partial_z G_0 \\ \partial_t G_1 &= -\eta G_1 - \delta(z-1)\partial_z G_1 + rz\partial_z G_0 \,. \end{aligned} \tag{3.21}$$

At this point it is instructive to compare the differential equations for the GGM to those of the PYM. The reaction (3.13), which describes the state change of the gene from G to G', is protein number dependent and leads to a change from a term $\sim rG_0$ to a term $\sim z\partial_z G_0$; otherwise, the equations remain unchanged for case (R). In case (A), the notable difference is the presence of the term $\sim \eta z G_1$ in the equation for G_0, which results from reaction (3.15). Note that the corresponding η-dependent term in the equation for G_1 does not depend on z: the symmetry between the equations is broken.

We solve the equations for case (A). Instead of solving for G_0 and G_1 separately we introduce the total generating function $G_T = G_0 + G_1$ with straightforward biological interpretation, allowing us to determine the mean value of the protein number and its fluctuations.

In case (A), the term $(z-1)$ can be placed in front in the equation for G_T, i.e.,

$$\partial_t G_T = (z-1)\Big(-\delta\partial_z G_T + \varepsilon(G_T - G_1) + \eta G_1\Big) \,. \tag{3.22}$$

One observes that a singularity arises at $z = 1$. The equality (3.22) is trivially satisfied at $z = 1$, because $G_T(1,t) = 1\ \forall t$. A stationary solution requires $\partial_t G_T(z,t) = 0\ \forall z, t$. For $z \neq 1$, this implies

$$-\delta \partial_z G_T + \varepsilon G_T + (\eta - \varepsilon) G_1 = 0\,. \tag{3.23}$$

Using Eq. (3.23), we can express G_1 and $\partial_z G_1$ in terms of $G_T(z,t)$. For the stationary solution, expression (3.21) can be rewritten in a second-order differential equation

$$\Big(\delta(\delta - z(r+\delta))\Big)\partial_z^2 g_T(z) + \Big(-\delta\varepsilon + \delta z\varepsilon + rz\eta - \eta\delta\Big)\partial_z g_T(z) + \eta\varepsilon g_T(z) = 0 \tag{3.24}$$

with $g_T(z) \equiv \lim_{t\to\infty} G_T(z,t)$. Substitution of $x = \delta(\delta - z(r+\delta))$ in Eq. (3.24) leads to the Kummer equation, familiar from the theory of ordinary differential equations,

$$x \frac{\partial^2}{\partial x^2} \tilde{g}_T(x) + (a + bx) \frac{\partial}{\partial x} \tilde{g}_T(x) + c\tilde{g}_T(x) = 0\,. \tag{3.25}$$

Here $a = (\varepsilon r + \delta\eta)/(r+\delta)^2$, $b = (\delta\varepsilon + r\eta)/(\delta^2(r+\delta)^2)$, $c = \eta\varepsilon/(\delta^2(r+\delta)^2)$ and $g_T(z) = \tilde{g}_T(x)$. Since $x = 0$ is a weak singular point of the differential equation (3.25), according to the method of Frobenius, a power series solution can be found around the singular point, i.e.,

$$\tilde{g}_T(x) = C_1 \sum_{n=0}^{\infty} a_n x^n + C_2 x^{1-a} \sum_{n=0}^{\infty} b_n x^n\,. \tag{3.26}$$

The substitution of the first term, i.e., the power series $\sum_{n=0}^{\infty} a_n x^n$, into Eq. (3.25), leads to the KummerM or the hypergeometric function $a_{0\,1}F_1(c/b, a; -bx)$. The second term corresponds to the KummerU function which can be discarded because the probability of n proteins does not tend to zero for $n \to \infty$. Consequently, the appropriate generating function $g_T(z)$ reads

$$g_T(z) = C_1'\,{}_1F_1(c/b, a; b\delta(z(r+\delta) - \delta))\,. \tag{3.27}$$

The coefficient C_1' is determined by the condition $lim_{z\to 1} g_T(z) = g_T(1) = 1$ due to the continuity of the hypergeometric function ${}_1F_1$, so that $C_1' = {}_1F_1(c/b, a; b\delta r)$. The asymptotic solution in all parameters can be written as

$$\begin{aligned} g_T(z) = {}_1F_1\Big(\frac{\eta\varepsilon}{\delta\varepsilon + r\eta}, \frac{\varepsilon r + \eta\delta}{(r+\delta)^2}; \frac{(\delta\varepsilon + r\eta)(z(r+\delta) - \delta)}{\delta(r+\delta)^2}\Big)\Big/ \\ {}_1F_1\Big(\frac{\eta\varepsilon}{\delta\varepsilon + r\eta}, \frac{\varepsilon r + \eta\delta}{(r+\delta)^2}; \frac{(\delta\varepsilon + r\eta)r}{\delta(r+\delta)^2}\Big)\,. \end{aligned} \tag{3.28}$$

From the total generating function (3.27) and Eq. (3.23), we can also separately derive the asymptotic solutions $g_0(z) = lim_{t\to\infty} G_0(z,t)$ and $g_1(z) = lim_{t\to\infty} G_1(z,t)$. The full expressions of the asymptotic solutions are

$$\begin{aligned} g_0(z) = C_1' \Big[{}_1F_1(c/b, a; b\delta(z(r+\delta)-\delta)) + \\ \frac{1}{\eta-\varepsilon}\Big(\varepsilon {}_1F_1(c/b, a; b\delta(z(r+\delta)-\delta)) - \\ \delta^2 c \frac{r+\delta}{a} {}_1F_1(c/b+1, a+1; b\delta(z(r+\delta)-\delta))\Big)\Big] \end{aligned} \tag{3.29}$$

$$\begin{aligned} g_1(z) = C_1' \frac{1}{\eta-\varepsilon}\Big[-\epsilon {}_1F_1(c/b, a; b\delta(z(r+\delta)-\delta)) + \\ \delta^2 c \frac{r+\delta}{a} {}_1F_1(c/b+1, a+1; b\delta(z(r+\delta)-\delta))\Big]. \end{aligned} \tag{3.30}$$

The probability for measuring n proteins, $p_{0,n} + p_{1,n}$, is obtained from $(1/n!)\partial_z^n g_T(z=0)$. We obtain an analytical expression for the stationary protein distribution $P(n)$

$$P(n) = \frac{1}{n!}\left(\frac{\delta\epsilon + r\eta}{\delta(r+\delta)}\right)^n \frac{\left(\frac{\eta\epsilon}{\delta\epsilon+r\eta}\right)_n}{\left(\frac{\eta r+\eta\delta}{(r+\delta)^2}\right)_n} \frac{{}_1F_1\left(\frac{\eta\epsilon}{\delta\epsilon+r\eta}+n, \frac{\epsilon r+\eta\delta}{(r+\delta)^2}+n; \frac{-\delta(\delta\epsilon+r\eta)}{\delta(r+\delta)^2}\right)}{{}_1F_1\left(\frac{\eta\epsilon}{\delta\epsilon+r\eta}, \frac{\epsilon r+\eta\delta}{(r+\delta)^2}; \frac{(\delta\epsilon+r\eta)r)}{\delta(r+\delta)^2}\right)} \tag{3.31}$$

with $(.)_n$ being the *Pochhammer symbol*, $(x)_n = x(x+1)\cdot ... \cdot (x+n-1)$.

The mean protein number E and its fluctuations V are easily obtained from the first and second derivative of the generating function $g_T(z)$ with respect to z, evaluated at $z=1$. In the activated case, the resulting expressions are given by

$$E = C_1' \frac{c\delta(r+\delta)}{a} {}_1F_1(c/b+1, a+1; b\delta r) \tag{3.32}$$

and

$$\begin{aligned} V = C_1' \frac{c(c/b+1)b\delta^2(r+\delta)^2}{a(a+1)} {}_1F_1(c/b+2, a+2; b\delta r) + \\ C_1' \frac{c\delta(r+\delta)}{a} {}_1F_1(c/b+1, a+1; b\delta r) - \\ \Big(C_1' \frac{c\delta(r+\delta)}{a} {}_1F_1(c/b+1, a+1; b\delta r)\Big)^2. \end{aligned} \tag{3.33}$$

We obtain the following asymptotic solution for $g_T(z) = g_0(z) + g_1(z)$:

$$\begin{aligned} g_T(z) = {}_1F_1\Big(\frac{\eta}{\delta}, \frac{(\varepsilon+\eta)r+\delta\eta}{(r+\delta)^2}; \frac{\varepsilon(z(r+\delta)-\delta))}{(r+\delta)^2}\Big)/ \\ {}_1F_1\Big(\frac{\eta}{\delta}, \frac{(\varepsilon+\eta)r+\delta\eta}{(r+\delta)^2}; \frac{\varepsilon r}{(r+\delta)^2}\Big) \end{aligned} \tag{3.34}$$

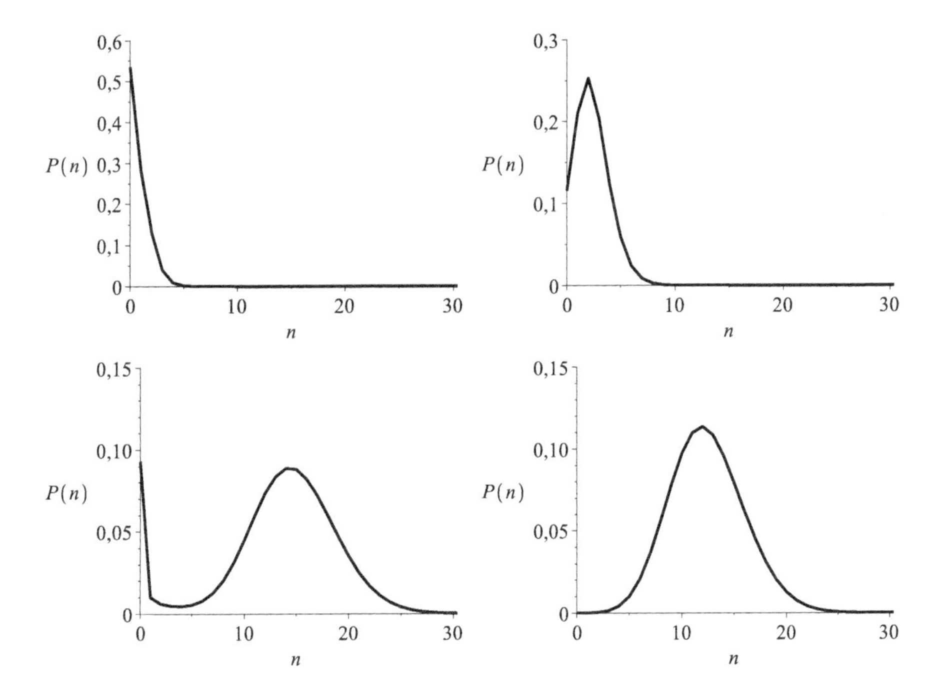

FIGURE 3.7 The probability distribution $P(n)$. Models and parameters from top left to bottom right: (A), $\varepsilon = 0.01$; $\eta = 0.1$; $r = 1.0$; $\delta = 0.1$; (A), $\varepsilon = 0.01$; $\eta = 0.1$; $r = 1.0$; $\delta = 0.04$; (A), $\varepsilon = 15.0$; $\eta = 0.1$; $r = 0.001$; $\delta = 1.0$; (A), $\varepsilon = 0.01$; $\eta = 0.1$; $r = 1.0$; $\delta = 0.008$. (Vandecan and Blossey, 2013). Reprinted with permission from the American Physical Society.

from which the probability distribution of the protein number, mean and variance can be derived. Figure 3.7 shows the resulting probability distributions for both the repressed and the activated gene. We find two generic scenarios depending on relative parameter values. If degradation dominates over protein production, the protein probability is peaked at $n = 0$ (shown for (A), but likewise for (R)). In case (A), an increase of protein production lets a peak at finite n grow out of the peak at $n = 0$. In the repressed case, a bimodal appears with a maximum at $n = 0$ and a peak at finite n; the same behavior is found for the activated model with inversed rates, i.e., for $\varepsilon > \eta$, as shown in Figure 3.7 (bottom, left). For still larger values of protein production, the protein probability becomes monomodal with a peak at finite protein number in both cases (R) and (A). These analytical results of the gene gate models predict accurately the results of the stochastic dynamics simulations of these simple gates. For the parameters $r = 1$, $\epsilon = 0.01$, $\eta = 0.1$ and $\delta = 0.001$, the formula (3.31) reduces to a Poisson-type distribution,

$$P(n) \approx \frac{e^{-\lambda}\lambda^n}{n!} \tag{3.35}$$

with a mean $\lambda = (\delta\epsilon + r\eta)/(\delta(r + \delta)) \approx 99$ proteins in steady state, in accordance with the simulations for the activated gene gate (Blossey, Cardelli and Phillips, 2006).

These results for this simple case show — aside from its astonishing technical difficulty for such a simple system — that already for the case of a simple gene the inclusion of stochastic effects is indeed crucial. In contrast to the deterministic model we developed before for the λ-phage, the simple gene does not contain any cooperative effects, which were so important for the λ-phage. Deterministic models of gene dynamics require cooperative interactions which then lead to nonlinearities in the dynamic equations and are necessary to obtain two states of gene. In the stochastic setting, one gets those two states — the bimodal distribution — without cooperativity.

Problem 5. Write down the extension of the gene gate model for two cases: (a) inclusion of cooperativity between transcription factors (dimerization); (b) include the translation step from mRNA into protein. What do the resulting master equations look like? How would you solve them?

References

Ackers G.K., A.D. Johnson and M.A. Shea
Quantitative model for gene regulation by λ phage repressor.
Proc. Natl. Acad. Sci (USA) 79, 1129-1133 (1982)

Blossey R., L. Cardelli and A. Phillips
A compositional approach to the stochastic dynamics of gene networks.
Transactions on Computational Systems Biology IV, in: Lecture Notes in Computer Science 3939, 99-122 (2006)

Blossey R., L. Cardelli and A. Phillips
Compositionality, stochasticity, and cooperativity in dynamic models of gene regulation.
HFSP Journal 2, 17-28 (2008)

Liu H., R. Zhang, W. Xiong, J. Guan, Z. Zhuang and S. Zhou
A comparative evaluation on prediction methods of nucleosome positioning.
Brief. Bioinf. 15, 1014-1027 (2014)

Peccoud J. and B. Ycart
Markovian modeling of gene-product synthesis.
Theo. Population Biol. 48, 222-234 (1995)

Ptashne M.
A Genetic Switch, Third Edition, Phage Lambda Revisited.
Cold Spring Harbor Press (2004)

Radmann-Liveja M. and O.J. Rando
Nucleosome positioning: how is it established, and why does it matter?
Dev. Biol. 339, 258-266 (2010)

Vandecan Y. and R. Blossey
Self-regulatory gene: an exact solution for the gene gate model.
Physical Review E 87, 042705 (2013)

Van der Hijden Th., J.J.F.A. van Vugt, C. Logie and J. van Noort
Sequence-based prediction of single nucleosome positioning and genome-wide nucleosome occupancy.
Proc. Natl. Acad. Sci. (USA) 109, E2514-E2522 (2012); Erratum: 110, 6240-

6242

Vanderlick T.K., L.E. Scriven and H.T. Davis
Solution of Percus's equation for the density of hard rods in an external field.
Phys. Rev. A 34, 5130-5131 (1986)

Further reading

Chevereau G., L. Palmeira, C. Thermes, A. Arneodo and C. Vaillant
Thermodynamics of intragenic nucleosome ordering.
Phys. Rev. Lett. 103, 188103 (2009)

Sheinman M. and H.-R. Chung
Conditions for positioning nucleosomes on DNA.
Phys. Rev. E 92, 022704 (2015)

Teif V.B., F. Erdel, D.A. Beshnova, Y. Vainshtein, J.-P. Mallm and K. Rippe
Taking into account nucleosomes for predicting gene expression.
Methods 62, 26-38 (2013)

Quintales L., E. Vázquez and F. Antequera
Comparative analysis of methods for genome-wide nucleosome cartography.
Brief. Bioinf. 16, 576-587 (2014)

CHAPTER 4

Histones and histone-acting enzymes

Histone proteins and post-translational modifications

So far we have entirely ignored the role played by the histone tails — the N-terminal ends of each of the histones which are largely unstructured and extend out from the nucleosome core particle. It seems that for the primary positioning of nucleosomes on DNA — and hence the lowest structural level of the chromatin fiber — the histone tails play only a negligible role. Polach, Lowary and Widom (2000) tested in their site exposure assays whether the presence or absence of the tails affects the accessibility of target sites occluded by nucleosomal DNA. Their finding was that the removal of the tails results in a significant but only small increase of the site-dependent equilibrium constants K_{eq}^c for site exposure by factors ranging from 1.5 to about 14.

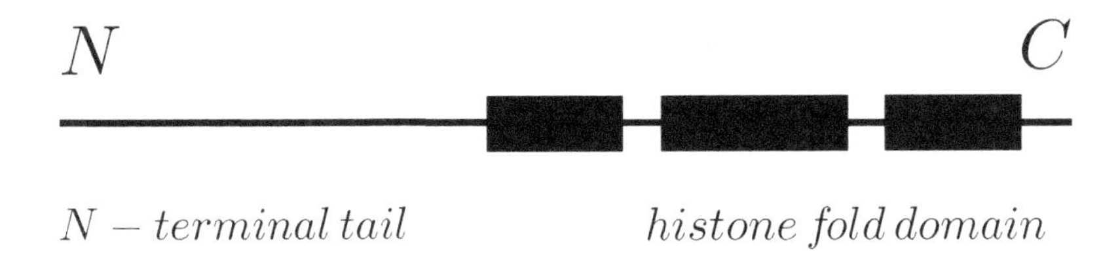

FIGURE 4.1 Histone build-up: tail and fold domain.

This change is not a really important effect — their role must obviously lie elsewhere.

Histone tails — for their build-up, see Figure 4.1 — and, as is currently more and more realized, also the histone cores are the substrate of a large number of *post-translational modifications*, PTMs for short. This notion refers to the covalent chemical modifications that the residues of the tails can undergo. These modifications are all described by a general reaction scheme which is detailed in the grey box below:

Histone tail modifications. All histone tail modifications follow a general enzymatic reaction scheme

$$\begin{array}{c} H + E \rightleftharpoons_{k_-}^{k_+} HE \rightarrow_{k_{cat}} H^* + E \\ \qquad\qquad\qquad\quad \downarrow_d \\ \qquad\qquad\qquad\quad H \end{array} \tag{4.1}$$

in which H denotes the histone tail (more specifically, a residue on the tail), E the enzyme, k_+ and k_- the binding and dissociation rates of the interaction. The enzymatic or catalytic reaction itself is denoted by k_{cat}, and it is performed by transferring a small molecule from the co-enzyme to the modified tail, H^*. The reaction denoted by d is the inverse reaction, which is not detailed. An exemplary modification pattern for histone H4 reads for the first 20 residues of the tail as

$$N-[S < p]GRG[K < ac]GG[K < ac]GLG[K < ac]GGA[K < ac]RHRK \tag{4.2}$$

where the notation reads [residue < modification].

The most prominent PTM in cellular signaling is phosphorylation p, in which a phosphate group is transferred via the consumption of ATP,

$$ATP \rightarrow ADP + Pi\,. \tag{4.3}$$

Phosphorylation of serine or threonine residues occurs on histones as well, but here it is not the most prominent modification. These are *acetylation*, ac, and *methylation*, m, which both target lysine residues. As the names say, in these cases there are acetyl and methyl groups transferred to specific residues along the histone amino acid chain. Their possible role in gene regulation was noted early on (Allfrey, Faulkner and Mirsky, 1964), even before the identification of the nucleosome.

Acetylation plays in fact two roles. The highly residue-specific acetylation is accompanied by an unspecific co-effect. The histone tails are positively charged and the acetyl group that is transferred carries a negative charge along, leading in sum to the reduction of the charge of the histone tail. This unspecific effect plays a role in the condensation/decondensation of chromatin.

A specific effect that has been invoked in this context is the *tail-bridging effect*, which shall serve here as an example of a condensation/decondensation effect. Mühlbacher, Schiessel and Holm (2006) studied this effect in a simple colloidal model in which the nucleosome is modeled as a negatively charged sphere, decorated by eight flexible polyelectrolyte chains of a given length. Each of the chains carries positive charges on each third monomer: all the parameters in the model were chosen to mimic closely the properties of real nucleosomes. Computer simulations of this model colloid system establish that the tail-bridging configurations, in which a tail from each of the colloid particles is closest to the tail of another colloid particle, lead to overall attractive forces between the particles, while the average interaction between non-bridging configuration is repulsive. This effect is modulated by the number of charges on the tails.

Another quantity the authors studied is the bond frequency, defined as the fraction of time the colloids are found in bridged configurations. This quantity is found to show an exponential decay if the Debye–Hückel screening length κ^{-1}, which is proportional to the square-root of the salt concentration in solution, fulfills the conditions $\kappa > 0.3\sigma^{-1}$ and $d > 9\sigma$, where $\sigma = 3.5\,\text{Å}$ is a chosen unit length, in terms of which the size of the colloid particle is given by $a = 15\sigma$, and d is the distance between the cores of the colloids. This behavior can be reasoned from the following argument the authors give. The monomers of the polyelectrolyte tails are adsorbed on either their "home" core or the "alien" core with an adsorption energy

$$\frac{\epsilon_{ads}}{k_B T} \approx \frac{V_{DH}^{mc}(0)}{k_B T} \tag{4.4}$$

where

$$\frac{V_{DH}^{mc}(r)}{k_B T} = Z\ell_B \frac{\exp(2\kappa a)}{1+\kappa a}\frac{\exp(-\kappa r)}{r} \tag{4.5}$$

is the Debye–Hückel screened electrostatic interaction between a monomer and the colloid core at a distance r, and ℓ_B is the Bjerrum length, of value 2σ in our case. Thus one has

$$\frac{\epsilon_{ads}}{k_B T} = \ell_B \frac{Z}{a(1+\kappa a)} \tag{4.6}$$

where Z is the smeared charge of the colloid. The cost of forming a bridge between two colloids can then be estimated from the number of monomers that need to be in between the colloids' cores. These are on the order of $\approx \lambda(d - \kappa^{-1})$ where $\lambda = f/\sigma$ is the line density of the bridge-forming tail, which is assumed to be stretched out. Here, f is the fraction of charged monomers. The probability of forming a bridge is thus given by the expression

$$p_{bridge} \propto \exp(-\epsilon_{ads}\lambda d/k_BT)\,, \tag{4.7}$$

a result which is confirmed by the simulations.

Let us return to the other types of modifications. Methylation, in contrast to acetylation, is charge neutral. Furthermore, while acetylation only adds a single acetyl group, methylation can add from one to three methyl groups per residue, which obviously increases the variability of these modifications. Further modifications are ADP ribosylation, ubiquitination and, most recently found, glycosylation (although this particular modification remains under debate).

A key contribution to the field has been the *histone code hypothesis* (Strahl and Allis, 2000) which proposes that the set of modifications on the histone tails constitutes a higher-order genetic code —an epigenetic code —of gene regulation and other processes. This proposal has clearly excited the chromatin research community, and meanwhile numerous correlations between histone tail modifications and cellular events (transcription, DNA repair, etc.) have been established, a general underlying principle however, which would allow a fairly simple mechanistic understanding of these modifications is still lacking. Progress is nevertheless being made, as we will see in Chapter 5 in the section on epigenetics.

Large-scale genomic data allow to gain a more general idea of the correlation between histone tail modifications and specific regulatory events. In a bioinformatics study, Karlić et al. (2010) could establish a link between specific histone acetylations and methylations and the expression levels of genes in CD4+ T-cells. They distinguished two different promoter types in terms of their CpG content (high or low) and could identify different abundances of specific modifications. When their spatial distribution in the vicinity of the transcription start sites of the genes is considered, shown in Figure 4.2, the result is strikingly similar to the typical positioning patterns found for the nucleosomes, in particular the acetylation signals.

Histone tail modifying enzymes

The complexity of the histone tail modifications is, as is evident from the description above, the result of the existence of a large variety of families of histone modifying enzymes which are commonly grouped as *histone writers, readers* and *erasers*. The level of complexity that resides even within one family of molecules is mind-boggling. Histone modifiers have one structural element in common, which is the recognition domain of the molecule. One of the most important families of histone enzymes is based on the *bromodomains* which are linked to histone tail acetylations.

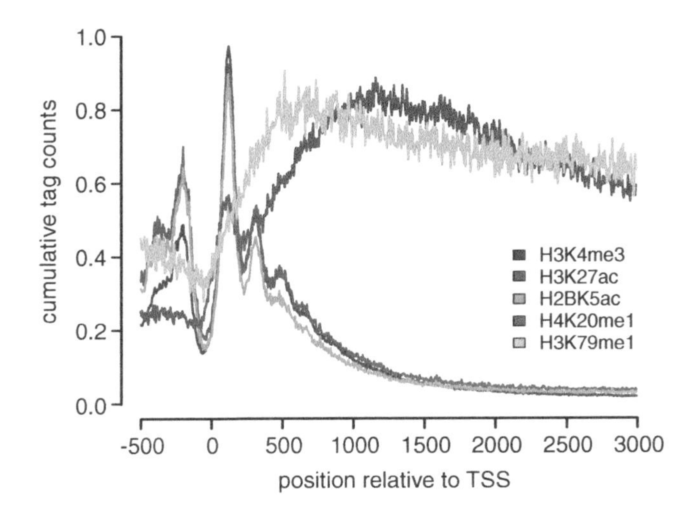

FIGURE 4.2 Spatial distributions of different histone tail modifications relative to the transcription start site. Reprinted with permission from Karlić et al. (2010). © PNAS.

Bromodomains provide key protein interfaces in gene expression mechanisms in eukaryotes (Filippakopoulos et al., 2012; Filippakopoulos and Knapp, 2014; Marmorstein and Zhou, 2014). Figure 4.3 displays a schematic phylogenetic tree of these domains according to current knowledge containing a totality of 56 species which can be grouped into eight families I-VIII based on sequence length and at least 30% sequence identity. In the figure, this tree has been reduced to the main branches, from which the specific bromodomains branch off as "twigs". The point this graph makes is to illustrate that, already within one specific recognition domain family, there is a complex family relationship.

We consider here GCN5 as an exemplary bromodomain which was studied in both human and yeast (Hudson et al., 2000; Owen et al., 2000). The GCN5 protein is ~439 amino acids long and contains a C-terminal bromodomain (BrD) which is composed of ~115 amino acids. It features the general characteristics of BrDs, i.e., a left-handed four-helical bundle (Z, A, B and C helices), and two ZA- and BC-loop regions with two (sometimes also three) small single-turn helices in the ZA-loop region. The active site of the GCN5-BrD is fairly narrow. Figure 4.4 shows a comparison of the GCN5 bromodomain structures for the two species (crystallographic data: PDB entries 1E61 for yGCN5 and 1F68 for hGCN5).

One method to demonstrate the specificity of the histone tail modifications on the tails is to consider the complex formation between the domain and its

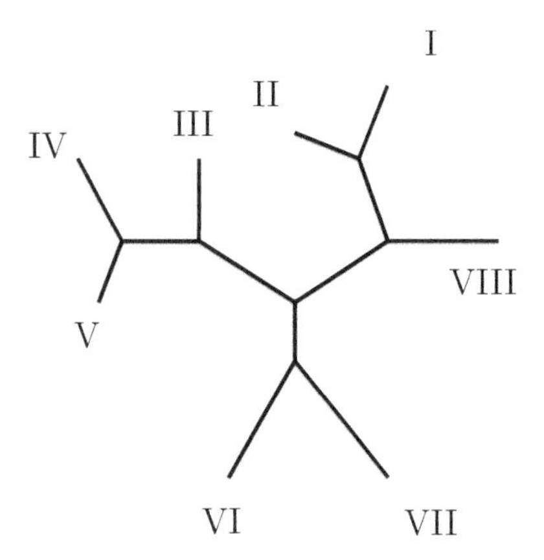

FIGURE 4.3 Sketch of only the backbone of the phylogenetic tree of bromodomains in humans, following Marmorstein and Zhou (2014) and (Fillipakopoulos et al., 2012).

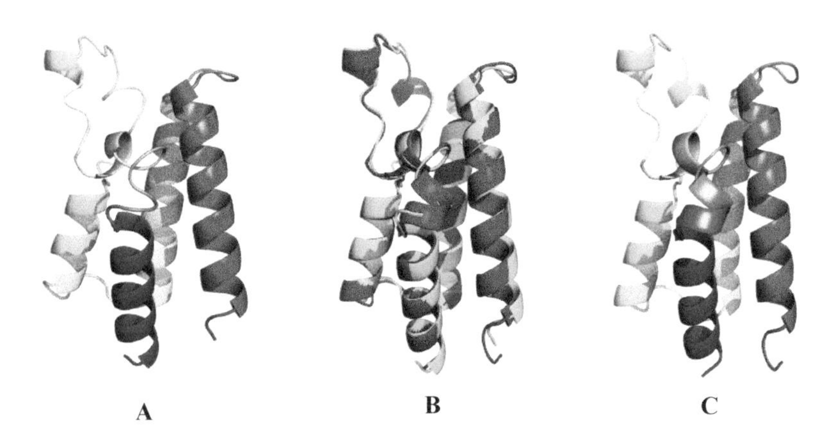

FIGURE 4.4 Ribbon-style representation of bromodomains. α-helices are shown as flat ribbons, loops as thin wires. (A) Yeast GCN5-BrD. (B) Structure-based alignment of yeast (purple) and human (green) GCN5-BrD with an RMSD of 0.8; small differences in the ZA- and BC-loops are observed. (C) Human GCN5. (Singh et al., 2015).

TABLE 4.1 Free energies of binding (FEB, acetylated vs non-acetylated) and dissociation free energies (DFE) for the histone tails H4 and H3 to the bromodomain GCN5 for yeast and human. Free energies of binding compare both acetylated (ac) and non-acetylated (nac) 15-mers at the indicated residue, while DFE values are given for the acetylated residue only (Singh et al., 2015). Free energy values are in kCal/mol.

ht	acK	Sequence	FEB	DFE
yH4	H4K5	GGSSGRG**K**GGKGLGK	-6.6/-5.4	24.10
	H4K8	SGRGKGG**K**GLGKGGA	-6.2/-6.1	16.37
	H4K12	KGGKGLG**K**GGAKRHR	-8.3/-5.6	24.37
	H4K16	GLGKGGA**K**RHRKILR	-8.2/-6.5	24.94
	H4K20	GGAKRHR**K**ILRDNIQ	-8.3/-5.4	27.04
hH4	H4K5	GGSSGRG**K**GGKGLGK	-6.2/-5.3	23.10
	H4K8	SGRGKGG**K**GLGKGGA	-6.1/-6.7	15.80
	H4K12	KGGKGLG**K**GGAKRH	-8.1/-5.4	25.89
	H4K16	GLGKGGA**K**RHRKVLR	-7.9/-6.0	24.23
	H4K20	GGAKRHR**K**VLRDNIQ	-8.3/-6.0	28.61
yH3	H3K4	STGGART**K**QTARKST	-6.8/-5.1	24.46
	H3K9	RTKQTAR**K**STGGKAP	-6.9/-6.6	17.53
	H3K14	ARKSTGG**K**APRKQLA	-8.4/-6.2	30.54
	H3K18	TGGKAPR**K**QLASKAA	-6.6/-6.2	24.51
	H3K27	LASKAAR**K**SAPSTGG	-6.5/-6.2	24.21
hH3	H3K4	STGGART**K**QTARKST	-6.3/-5.9	26.32
	H3K9	RTKQTAR**K**STGGKAP	-6.5/-5.1	16.46
	H3K14	ARKSTGG**K**APRKQLA	-5.8/-5.8	29.80
	H3K18	TGGKAPR**K**QLASKAA	-6.7/-6.2	22.05
	H3K27	LASKAAR**K**SAPSTGG	-6.5/-6.3	23.79

ligand, modified at different residues. Singh et al. (2015) adopted an approach combining molecular dynamics (MD), peptide docking and umbrella sampling. Histone tail peptides consisting of 15 amino acids with a central acetyl-marked lysine flanked by 7 amino acids on each side in the form (7aa-acK-7aa) were constructed representing five different acetylation marks taken at different positions along each of the histone tails H3 (lysine K4, K9, K14, K18 and K27) and H4 (lysine K5, K8, K12, K16 and K20). The specific sequences as well as the resulting dissociation free energies are shown in Table 4.1 and plotted in Figure 4.5.

For the H4 tail, a distinctly lower value of the acetylated residue K8 is clearly discernible. This result underscores that the positioning of the acetyl-mark on residue K8 is associated with a preferred release of the bromodomain from the tail. The case of H3 is distinct from the previous case in that we can see that for residue K9 the DFE is equally lowered (as for the case of K8 in H4) while

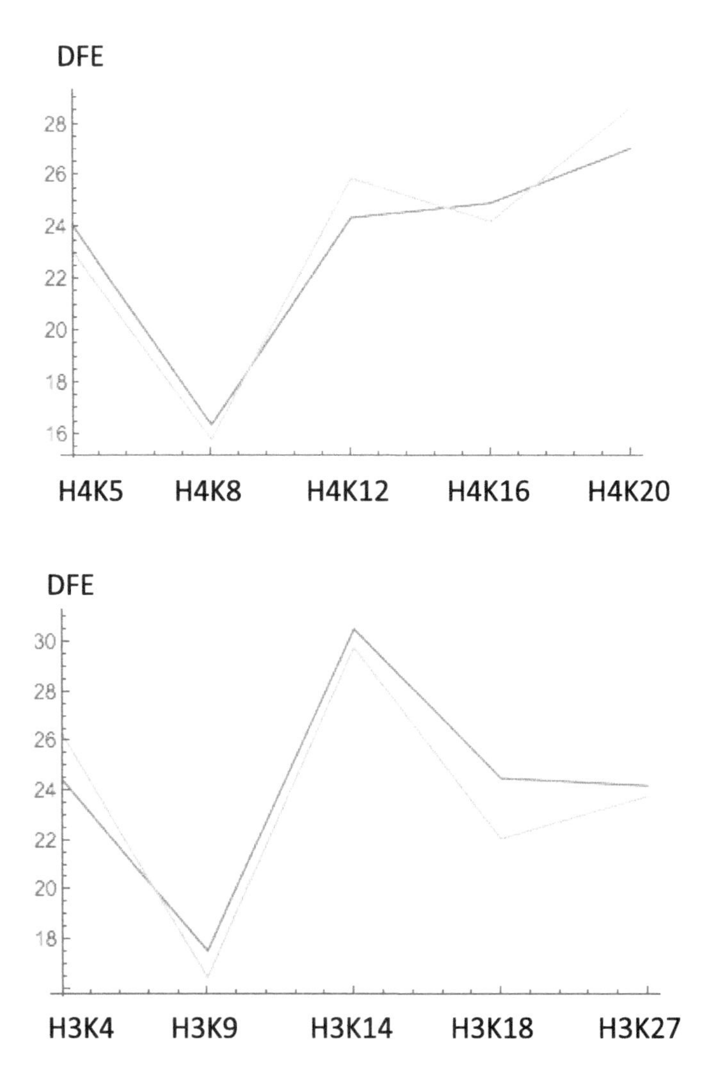

FIGURE 4.5 Dissociation free energies (in kCal/mol) as determined from the pulling protocol, as taken from Table 4.1. On the abscissa the modified histone amino acids are placed on arbitrary integer positions; the drawn line is a guide to the eye. Top: H4; bottom: H3. The lighter curve is the yeast data (Singh et al., 2015).

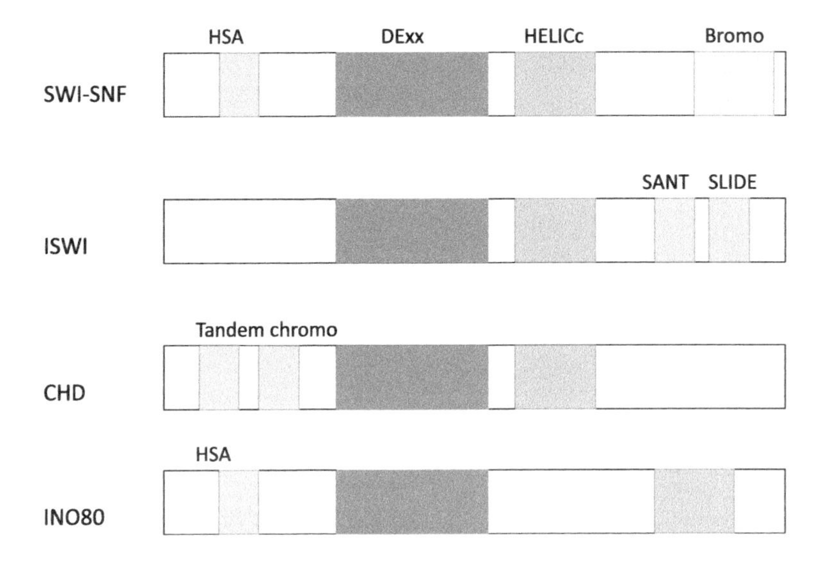

FIGURE 4.6 Chromatin remodelers are enzyme-motor families defined by their ATPase. All remodelers contain a SWI2/SNF2-family ATPase subunit characterized by a domain split in two parts: DExx (dark grey) and HELICc (lighter grey). What distinguishes each family are the unique domains alongside the ATPase domain. Each family is further defined by flanking domains: e.g., bromodomains for the SWI/SNF family and the SANT-SLIDE module for the ISWI family. Redrawn from Clapier and Cairns (2009).

the value for acetylation of K14 is in fact increased. We will interpret these findings in the context of the regulation of gene activation in the following chapter.

Chromatin remodelers

We now turn from the histone tail modification writers to a particular set of readers: the *chromatin remodelers.* These large complexes have not only enzymatic capacities — they are, in fact, molecular motors that act on nucleosomes. The ATPase unit of chromatin remodelers is related to helicases from the SNF2 family which are molecular motors that open DNA double strands; remodelers have evolved to unpeel the DNA from the histone octamer. Figure 4.6 displays the schematics of the sequence of the four known remodeler families. Alongside the ATPase, accessory domains interact with histone tails and their modifications: the bromodomain, in the case of remodelers from the

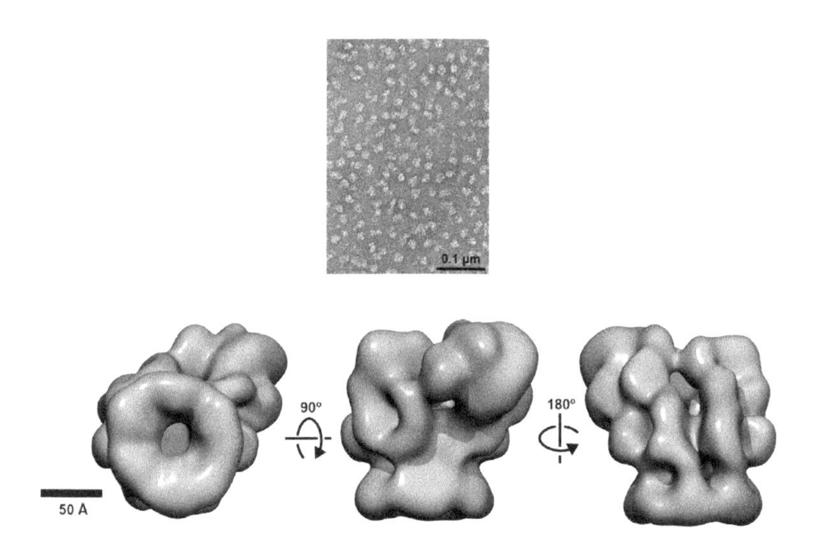

FIGURE 4.7 Cryo-EM reconstructions of the remodeler SWR1, a member of the INO80 family. Shown are the captured structures (top) and resulting reconstruction (bottom) at a resolution of 28 Å. Reprinted with permission from Nguyen et al. (2013), © Elsevier.

SWI-SNF family, and with DNA. Given that the molecular structures of these complexes are huge, they are only recently being resolved with cryoelectron microscopy; see Figure 4.7 for the example of SWR1, a member of the INO80 family of remodelers.

In the following, we will discuss two exemplary cases from these families. The first is the remodeler RSC, which belongs to the SWI-SNF family. The second is ISWI from the family with the same name.

Helicases and remodelers. A helicase is an enyzme and molecular motor which opens up the double-stranded DNA in order to provide access to the sequence. In this sense it is not astonishing that chromatin remodelers are a related enzyme class: their action is to remove the DNA double strand from the histone octamer.

RSC. Remodelers often carry acronyms as their names which can refer to very specific biological processes, but RSC does not: it stands for *R*emodels the *S*tructure of *C*hromatin.

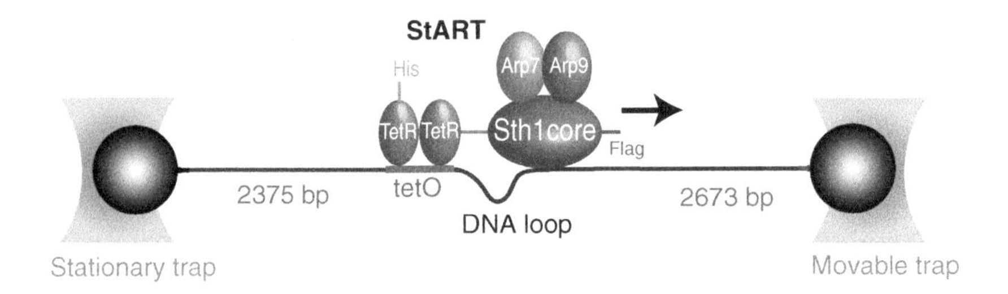

FIGURE 4.8 Construction of the RSC remodeling assay. The remodeler is held in place on DNA by binding to two transcription factors TetR. The DNA segment is held between two optical tweezers. The remodeler pulls the two beads toward each other by creating a looped DNA. Reprinted with permission from Sirinakis et al. (2011) © Wiley.

RSC

The structure of the remodeler RSC has been determined by a cryo-electron microscopic reconstruction of the complex (Leschziner et al., 2007), work followed up by (Chaban et al., 2008) in a similar fashion as shown above for SWR1. The reconstruction showed that RSC can take up the whole nucleosomal complex and close up around it. How RSC actually "remodels", i.e., displaces or removes a nucleosome, cannot be learned from structure alone, and single-molecule-type assays of fully reconstituted remodeling assays have been unsuccessful to this date to clarify this issue. The full complex has simply proven to be too difficult to handle at present. However, this does not mean that one cannot access the properties of the remodelers by such types of assays — one just needs to modify them in a proper way.

Here we describe results from Sirinakis et al. (2011), who employed an optical tweezer setup in order to establish basic characteristics of RSC. Instead of working with reconstituted remodelers, they built a clever construct, based on DNA strands of about 2 kbp length. On this they placed a construct made from the core ATPase domain of RSC, Sth1, stabilized by two actin-related proteins; see the drawing in Figure 4.8. This "reduced" remodeler was then linked to TetR transcription factors which in turn are bound to corresponding promoter sequences on DNA.

Upon fueling the remodeler with ATP, the authors observed the generation of DNA loops which are established from an analysis of the forces acting on the optically trapped beads. ATP concentration affects remodeler speeds and *processivity* in different ways: while an increase of ATP widens the loop size distribution slightly, it tends to sharpen the distribution of observed speeds,

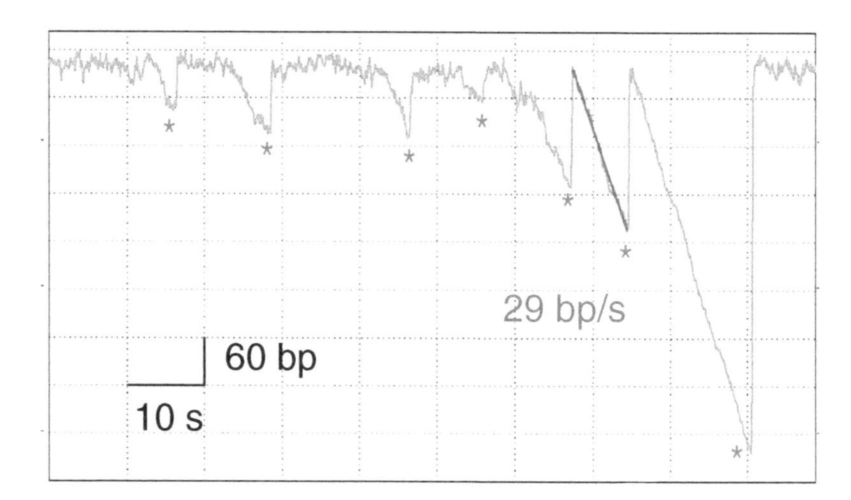

FIGURE 4.9 Force vs. time traces of the remodeler action on the beads which are identified with DNA loops. Speeds and distances are calculated from linear fits as indicated. Reprinted with permission from Sirinakis et al. (2011) © Wiley.

while not much affecting the average speed, which is found to be about 25 bp/s; see Figure 4.9. The processivity lies at about 35 bp, with a step size of about 2 bp. Interestingly, the maximum applied force against the motor lies at around 30 pN, which makes RSC one of the strongest molecular motors found. This maximal force is clearly sufficient to disrupt the wrapped DNA from the histone octamer. Figure 4.10 shows the speed and processivity distributions for the remodeler as functions of ATP concentration.

Processivity. In molecular biology and biochemistry, processivity is the capacity of an ezyme to catalyze consecutive reactions without releasing its substrate; in our case, the remodeler acts without falling off the nucleosome.

This in vitro experiment, despite its fairly constructive character, which might impede a direct transfer of its interpretation to the in vivo case, shows very clearly that under the fueling by ATP, the remodeler is capable of exerting very strong forces on DNA. It is fair to say that the nucleosome in some sense is simply used by the remodeler to anchor itself, and to move the DNA relative to this anchoring point. It would in this way be better not to say that remodelers shift nucleosomes, but rather move DNA away from them

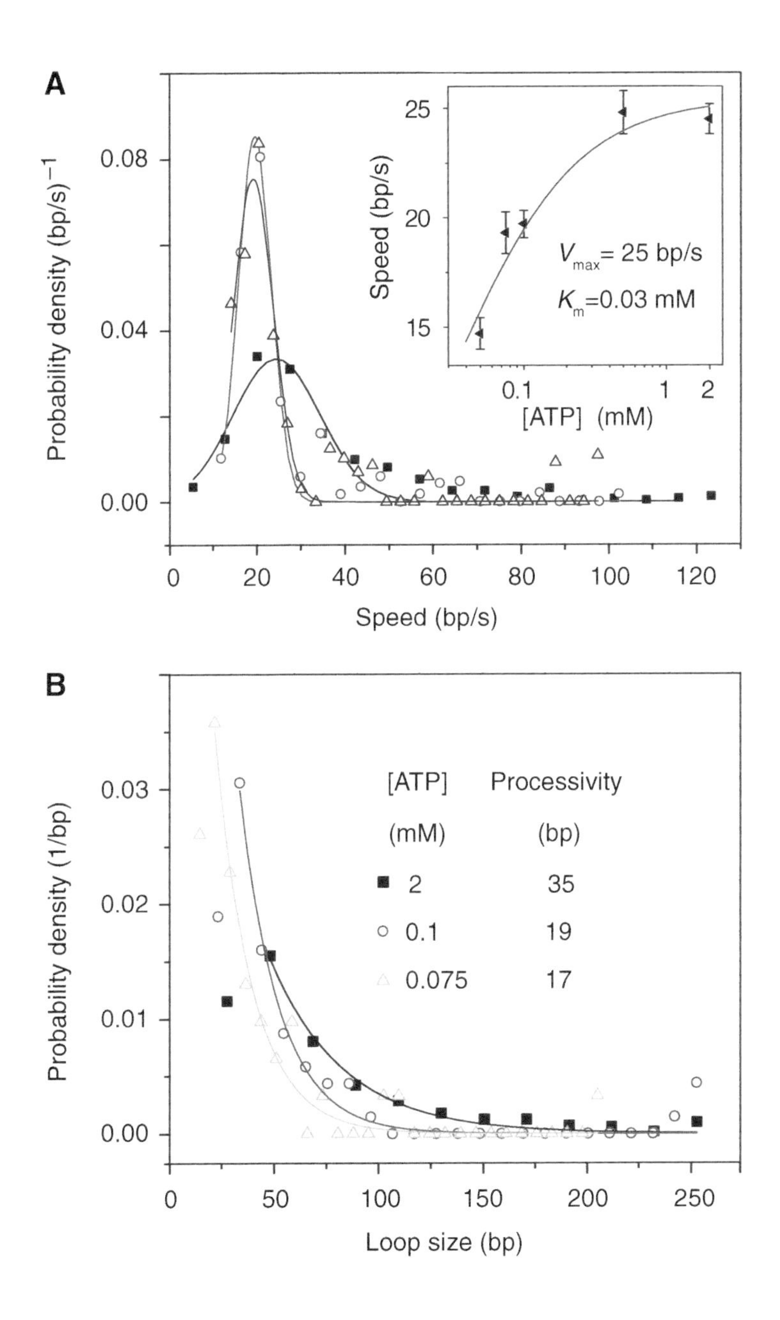

FIGURE 4.10 Speed and processivity distributions of the RSC remodeler. The first are Gaussian, the latter exponential. Reprinted with permission from Sirinakis et al. (2011) © Wiley.

(although obviously the outcome of the interaction amounts to the same).

ISWI

The ISWI family of remodelers engages mainly in the repression of genes. The remodeler contains besides the ATPase domain in particular a domain extending to and interacting with DNA, the SLIDE domain. Its action of DNA is a bit like that of a sailor pulling in a rope and moving it around a pole. This domain is an essential element for the functioning of the remodeler, as it is involved in both the motion of DNA into the nucleosome, as well as providing directionality for the motion of DNA. If its action is impaired, the ATPase alone can still move nucleosomal DNA, but will stall after about 10 bp. This finding also shows a major difference between the action of RSC, as a member of the SWI-SNF familly, which does not require extra-nucleosomal (linker) DNA for its action, while ISW2 does. This reflects the biological purpose of these machines: while RSC is involved mainly in activation of gene transcription, ISW2 is needed for positioning of nucleosomes to prepare for gene repression.

Nucleosome positioning with remodelers

In this section we look at the role remodelers play in the positioning of nucleosomes. Meanwhile we have seen a number of effects that influence nucleosome positioning along the chromatin fiber, such as DNA sequence, both with respect to bending and sequence content, and all of this combined with statistical effects. Histone tail modifications also appear to be involved, but here the picture is less clear — based on the analysis of high-throughput data we can only establish correlations, but we do not understand the underlying mechanism.

Before going up to the genome level, it is again interesting to look at a "single molecule" assay in which the positioning of a single nucleosome has been studied. This is work performed by G.J. Narlikar and collaborators (Yang et al., 2006; Blosser et al., 2009; Racki et al., 2009), which is based on a variant of ISWI called ACF. The setup of the assay is shown in Figure 4.11. A nucleosome is placed on a DNA segment, and it is initially positioned at one extremity of the sequence, as the latter has been constructed by including a strong positioning sequence at this end. The initial position of the nucleosome is thus well known. Both the DNA sequence and the nucleosome are labelled by chromophores, indicated by differently colored circles in the sketch. The energy transfer between the labels thus allows us to elucidate the position of the nucleosome on the DNA. By studying the binding kinetics of ACF on the nucleosome, one can first find out that indeed ACF is binding to the substrate in the form of a dimer.

FRET. The abbreviation stands for *Förster resonance energy transfer*, alternatively *fluorescence resonance energy transfer*. It describes the energy transfer between two light-sensitive molecules (chromophores). A donor molecule which is initially in an electronic excited state may transfer the energy to an acceptor chromophore via a radiationless dipole–dipole coupling. The efficiency of energy transfer is inversely proportional to the sixth power of the distance between donors and acceptors. This dependence makes FRET very sensitive to small distance changes in the position of the molecules.

The second step then is to elucidate the repositioning, which can be done by tracing out the FRET signal. ACF centers the nucleosome on the DNA, thereby overruling the built-in sequence preference to position the nucleosome. The effect is due to the SLIDE domain: SLIDE senses linker length and coordinates the action of the two remodelers in such a way that the dimer prefers to move the nucleosome toward the longer length, resulting in this "centering"-effect.

A mathematical model for nucleosome positioning by ACF

We now discuss a master equation approach to nucleosome positioning motivated by the experimental work by G.J. Narlikar and her collaborators. The advantage of a model obviously is that one can try out a number of different set-ups to see how robust the situation actually is. Here, in particular, we will distinguish between models of different levels of (experimental) detail put into the model. In particular, we will consider the case of an "effective" one-motor model and a two-motor model in which the coordination of the two motors becomes an issue. Conformational changes between the ATPase and the extranucleosomal DNA binding domain are expected to pull in the linker DNA from either side (entry and exit site), resulting in the idea that one motor is translocating to the "left" and the other to the "right". The idea of the alternating action between the two ACF motors and the dependence of the ATP-hydrolysis rate on the DNA flanking length then leads to a competition between back and forward movement of the nucleosome, which in this simple first model is translated into the length dependence of the transition rates. A schematic drawing of the situation we address is given in Figure 4.12.

As a starting point for our mathematical analysis we assume that a nucleosome in solution can be in one of three states. The first state corresponds to a free nucleosome, not bound to the remodeler ($N + R$). In the second state, the

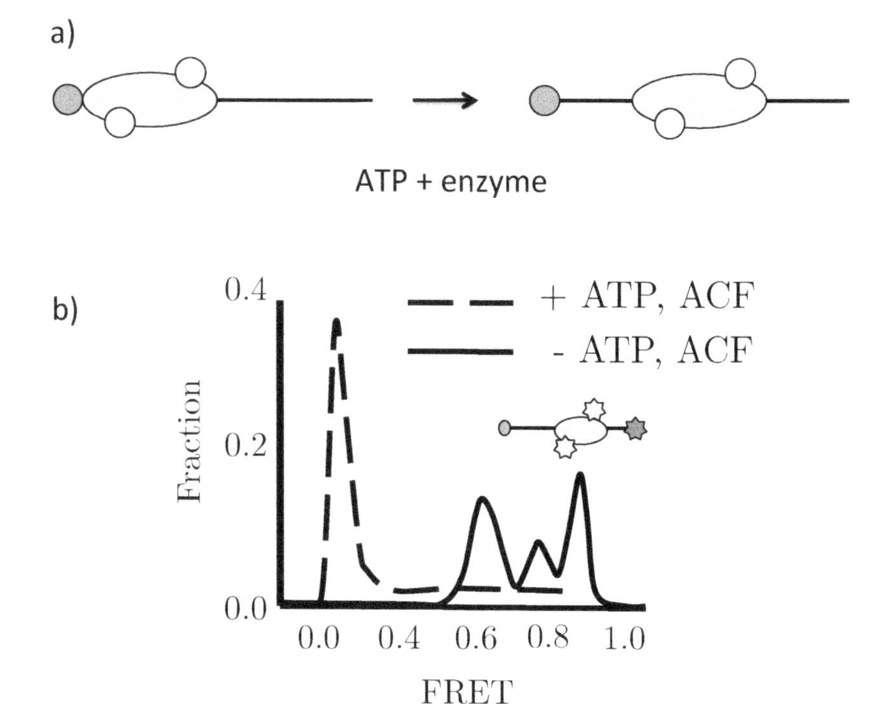

FIGURE 4.11 (a) Schematic drawing of the FRET-based nucleosome remodelling assay, following Racki et al. (2009). (b) Details of the FRET experiment. The initial three FRET signals shown in the drawn line for the end-positioned nucleosome are depleted after addition of ATP and the remodeler. Adapted from Blosser et al. (2009).

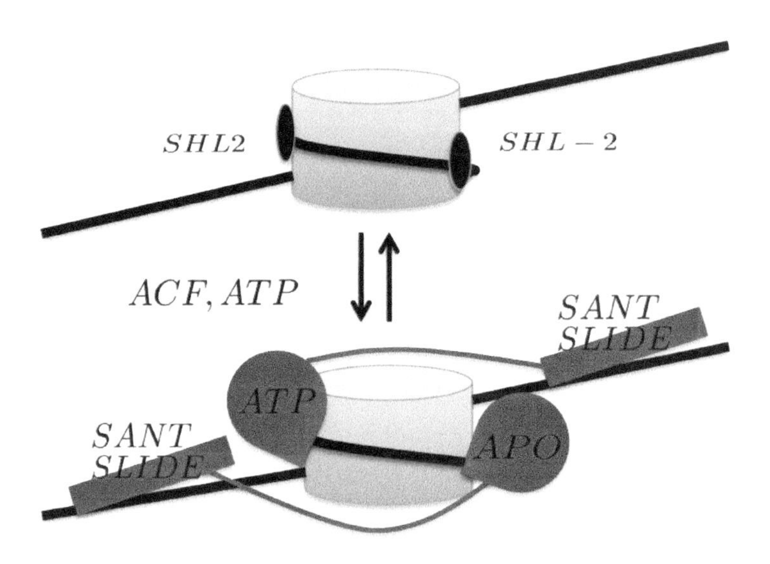

FIGURE 4.12 Schematic drawing of the positioning of the remodeler ACF as a dimer on a nucleosome (Vandecan and Blossey, 2012). Reprinted with permission from the American Physical Society.

nucleosome is bound to the remodeler, forming a nucleosome–remodeler complex (NR). Finally, the third state refers to partially loosened DNA around the nucleosome due to the remodelers' consumption of one ATP (NR^*). In this loosened state, successive intermediates are generated by the successive movement of 13 ± 3 bp of DNA across the histone octamer, thereby consuming one ATP molecule per remodeling event. The three-state model states with appropriate rates are schematically depicted in Figure 4.13.

As can be seen in Figure 4.13, we consider this model with seven different intermediate states along the DNA. The seven repositioned nucleosomal states are ($0N78, 13N65, 26N52, 39N39, 52N26, 65N13$ and $78N0$) ($n = 1, \cdots, 7$), corresponding to the experimentally observed states. We represent the on-rate from $N + R$ to NR as k_c, the off-rate from NR and NR^* to $N + R$ as k_c', the rate from NR to NR^* as k_I (activation step) and the extranucleosomal DNA dependent remodeling rates in the direction $n \to n + 1$ ($n \to n - 1$) as $k_n(k_n')$. Further, the $P_k(n, t)$, $k = 1, 2, 3$, are, respectively, the probabilities of the nucleosome–remodeler complex to occupy the states $N + R$, NR and NR^* with n referring to the position of the nucleosome along the DNA strand. We can then set up a master equation for the transitions between these states

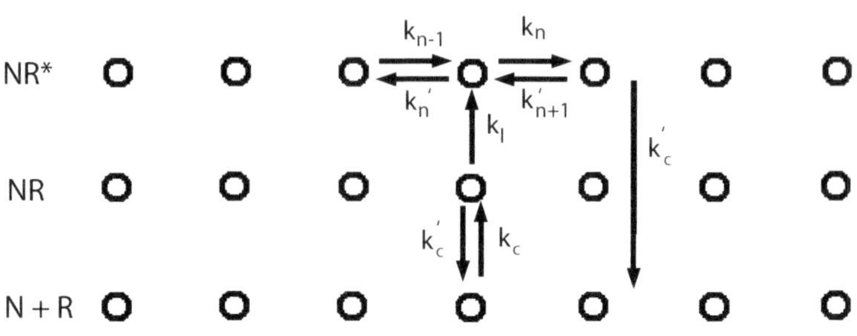

FIGURE 4.13 The small circles of each row represent the nucleosomal intermediates with different extranucleosomal length (n = 1.7). Each row itself refers to $N + R$, NR and NR^*, respectively. The proper rates between the several states are indicated by arrows (Vandecan and Blossey, 2012). Reprinted with permission from the American Physical Society.

as

$$\begin{aligned} \frac{\partial P_1(n,t)}{\partial t} &= -k_c P_1(n,t) + k_c' P_2(n,t) + k_c' P_3(n,t) \\ \frac{\partial P_2(n,t)}{\partial t} &= k_c P_1(n,t) - (k_I + k_c') P_2(n,t) \\ \frac{\partial P_3(n,t)}{\partial t} &= k_I P_2(n,t) - (k_c' + k_n' + k_n) P_3(n,t) + \\ & \quad k_{n-1} P_3(n-1,t) + k_{n+1}' P_3(n+1,t) \,. \end{aligned} \tag{4.8}$$

According to Michealis–Menten kinetics, we consider the following enzymatic reaction scheme for the DNA translocation step

$$N_n R^* + ATP \rightleftharpoons N_n R^* - ATP \rightarrow N_{n\pm 1} R^* + ADP + P \,. \tag{4.9}$$

Here $N_n R^*$ represents the nucleosome–remodeler complex where the nucleosome is in the loosened intermediate state n without ATP bound, and $N_n R^* - ATP$ reflects the same complex with ATP bound. The reaction rate r of the formation of the product, which here is the remodeling step, is given by

$$r = V_{max} \frac{[ATP]}{[ATP] + K_m} \tag{4.10}$$

where V_{max} is the maximal speed of the reaction, $[ATP]$ the ATP concentration and K_m the Michaelis constant.

Using the Arrhenius equation, the maximal reaction speed V_{max} is proportional to $e^{-E_a/kT}$ where E_a is the activation energy. If the remodeler concentration is saturated, this concentration must not be taken into account. The experiments show that the activation energy varies linearly with the DNA flanking length up to 60 bp for the remodeler ACF. As a consequence, the transition rates are linker-length dependent. From 60 bp up to $\approx$ 100 bp, the transition rates are assumed to be constant, but are expected to decrease to zero at a DNA flanking length of $\approx$ 160 bp due to loss of contact of the remodeler domain at the SHL-2 or SHL2 locations on the nucleosome (see Figure 4.12). The same holds for values of the DNA flanking length lower than 20 bp. Analytically, the transition rate dependence then takes the following form:

$$k_n = \begin{cases} 0, & 0 < \ell < \ell_{min} \\ k_0 e^{a\ell}, & \ell_{min} \leq \ell \leq \ell_{max} \\ k_0 e^{a\ell_{max}}, & \ell_{max} < \ell \leq \ell_c \,. \end{cases} \tag{4.11}$$

The experimentally estimated parameter values for an ATP concentration of 10 μM (from gel mobility shift experiments) are $k_0 = 0.057/\text{min}$, $a = 0.077\,\text{bps}^{-1}$, $\ell_{min} = 20\,\text{bps}$, $\ell_{max} = 60\,\text{bps}$ and $\ell_c = 100$ bps. Concerning the remaining rates for an ATP concentration of 10 μM, we choose $k_c = 12/\text{min}$, $k_c' = 8/\text{min}$ and $k_I = 27/\text{min}$; see Table 4.2.

From the experimental data of the gel mobility shift experiment, Eq. (4.10) allows us to compute k_n for different ATP concentrations. It suffices to calculate $k_0([\text{ATP}])$ because the parameter a is independent of the ATP-concentration. The Michealis–Menten constant is taken to be 50 μM. For comparison with the extended two-motor model discussed in the following we restrict our calculations to the initial condition $0N78$ and set $k_c' = 1/\text{min}$, which corresponds to a high processivity.

TABLE 4.2 The values of the model parameter k_0 as a function of the ATP concentration (Vandecan and Blossey, 2012). Reprinted with permission from the American Physical Society.

$[ATP](\mu\text{M})$	$k_0(min^{-1})$
2	0.013
10	0.057
20	0.098
200	0.27
2000	0.33

The master equation is a system of coupled linear partial differential equations of first order which can be written in the form

$$\frac{\mathrm{d}}{\mathrm{dt}}\vec{P} = \mathbf{M}\vec{P} \tag{4.12}$$

with $\vec{P} = \left[P_k(n,t)\right]^{\mathrm{T}}$ with $(k,n) = (3,7)$ a (21×1)-vector and $\mathbf{M}$ a (21×21)-rate matrix for $n = 7$. Substitution of $\vec{P} = \mathbf{C}\vec{P}_N$ in Eq. (4.12), with $\mathbf{C}$ the invertible matrix of eigenvectors of $\mathbf{M}$, transforms the system into a set of uncoupled linear partial differential equations, i.e.,

$$\frac{\mathrm{d}}{\mathrm{dt}}\vec{P}_N = \mathbf{D}\vec{P}_N \tag{4.13}$$

with $\mathbf{D}$ the diagonal matrix of eigenvalues λ_i for $i = k \cdot n = 1..21$ and $\vec{P}_N$ the vector of normal modes. The general solution is then obtained in terms of a linear combination of normal modes,

$$P_i(t) = \sum_{j=1}^{N_o} u_{ij} c_j e^{\lambda_j t}\,, \tag{4.14}$$

with $P_i(t)$ the component i of the vector $\vec{P}$, N_o the length of the vector $\vec{P}$, u_{ij} the components of the matrix $\mathbf{C}$ and c_j the constants referring to the initial condition.

The effective translocation steps with rate k_n (k'_n) in Figure 4.13 consist of translocation and ATP-loading steps of the two motors. We can thus introduce a two-motor model simply by splitting the previous nucleosome–remodeler state NR^* ($0N78$, $13N65$, $\cdots$) into four explicit motor states, i.e., (ATP,0), (ADP,0), (0,ATP) and (0,ADP). The (ATP,0) and (ADP,0) states reflect that the adenosine-tri(di)phosphate is bound at the ATPase unit of motor 1 while the other motor is in the apostate (unloaded state), and vice versa for the other two states. Considering transitions between these different motor states, we let the motors communicate as shown in the schematic graph in Figure 4.14.

Hydrolysis of ATP (ATP $\rightarrow$ ADP) with the subsequent translocation step is depicted by large black arrows with rate k_n or k'_n. We clearly see that the ATP hydrolysis at the ATPase unit of motor 1 translocates in the opposite direction as compared to motor 2. In order to include the dependence of the rate of ATP hydrolysis with the subsequent translocation step on the extranucleosomal DNA length, we again apply Eq. (4.11), but with a modified k_0 value. The parameter a reflects the intrinsic flanking-length dependence and remains unchanged.

The arrows with rate k_p refer to the ATP-loading phase, i.e., the dissociation of ADP with the binding of a new ATP molecule to the ATPase unit of motor

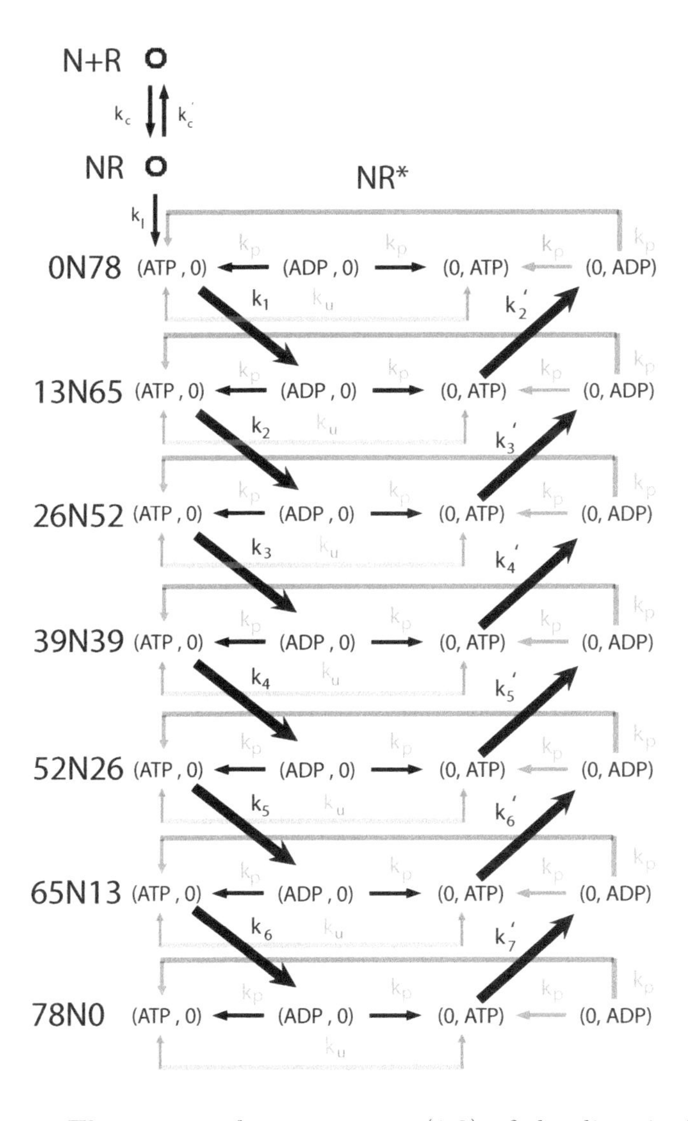

FIGURE 4.14 The presumed motor states (1,2) of the dimeric ACF remodeler with their transitions. The big, black arrows denoted by k_n and k'_n reflect the translocation step using ATP hydrolysis, i.e., the transition from (ATP,0) to (ADP,0) state for motor 1, analogously for motor 2, but then with the reverse translocation step. The ATP-loading step is split into two distinct transitions. The first type of transitions is denoted by the arrows with rate k_p, which represent the reloading from an ADP to an ATP state after the translocation step. The second type is depicted by arrows with rate k_u, which reflect the unbinding of the ATP molecule at motor 1 and binding of an ATP molecule to motor 2, or vice versa (Vandecan and Blossey, 2012). Reprinted with permission from the American Physical Society.

1 or 2. We include also transitions between (ATP,0) and (0,ATP) (arrows with rate k_u), which represent the unbinding of a non-hydrolyzed ATP molecule belonging to the ATPase unit of motor 1 (2) and subsequent binding of a ATP molecule at the ATPase unit of motor 2 (1). These stages reflect ATP loading, and are presumably related to a kinetic pausing of the remodeler observed in the experiments.

In view of a highly processive motor (with a low k'_c value) and in order to gain insight into the effect of ATP loading and translocation separately, it suffices to restrict our description to the active remodeling phase together with the $N+R$ and NR states of the end-positioned nucleosome $0N78$. Since we neglect the dissociation of the remodeler in the NR^* states, the $N+R$ and NR state of the $0N78$ state will only induce a time delay. These assumptions reduce the number of possible states to 28 states. The extended master equation of this two-motor model can then be written as

$$\frac{\partial P_1(t)}{\partial t} = -k_c P_1(t) + k'_c P_2(t) \tag{4.15}$$

$$\frac{\partial P_2(t)}{\partial t} = k_c P_1(t) - (k_I + k'_c) P_2(t) \tag{4.16}$$

$$\frac{\partial P_3(n,1,t)}{\partial t} = -(k_n + k_u) P_3(n,1,t) + k_p\big(P_3(n,2,t) + P_3(n,4,t)\big) + k_u P_3(n,3,t) + k_I P_2(t)\delta_{n,1} \tag{4.17}$$

$$\frac{\partial P_3(n,2,t)}{\partial t} = -2k_p P_3(n,2,t) + k_{n-1} P_3(n-1,1,t) \quad (n \neq 1) \tag{4.18}$$

$$\frac{\partial P_3(n,3,t)}{\partial t} = -(k'_n + k_u) P_3(n,3,t) + k_p\big(P_3(n,2,t) + P_3(n,4,t)\big) + k_u P_3(n,1,t) \tag{4.19}$$

$$\frac{\partial P_3(n,4,t)}{\partial t} = -2k_p P_3(n,4,t) + k'_{n+1} P_3(n+1,4,t) \quad (n \neq 7) \tag{4.20}$$

with the initial condition $P_1(t=0) = 1$. $P_1(t)$ and $P_2(t)$ represent the $N+R$ and the NR states of the end-positioned nucleosome, respectively. $P_3(n, M, t)$ reflects the loosened state NR^* with $n = 1, \cdots, 7$ being the previously discussed intermediates, and $M = 1, \cdots, 4$ the four motor states, i.e., (ATP,0), (ADP,0), (0,ATP) and (0,ADP), in that order. The Kronecker symbol $\delta_{n,1} = 1$ when $n = 1$ and zero otherwise. Furthermore, $P_3(1,2,t) = P_3(7,4,t) = 0$ due to the initial condition $P_1(t=0) = 1$. This extended master equation can be solved with the same mathematical technique as discussed above.

Results and discussion: the effective one-motor model vs the two-motor model

Figure 4.15 shows the probability distributions to find the nucleosome over time along the DNA, for two different initial conditions, in the one-motor model. Irrespective of the initial condition, the steady-state probability $P(n = 4, t)\,(39N39)$ reaches nearly 60%, indicating that the remodeler produces mainly center-positioned nucleosomes. Figure 4.16 displays the motor or nucleosome sliding velocity and the diffusion constant for the one-motor model. The motor velocity v along the DNA chain is computed through

$$v = \frac{\mathrm{d}}{\mathrm{dt}}\left(\sum_{n=1}^{7} P(n,t)x_n\right). \tag{4.21}$$

Here, x_n is the nucleosome–remodeler position with respect to the DNA. For end-positioned nucleosomes the nucleosome sliding velocity reaches a maximum of ≈ 0.5 bp/s at $t \approx 20$ s (at an ATP concentration of 10 μM) to converge to zero velocity when the nucleosome is center-positioned with respect to the DNA chain (see Figure 4.16a). The steady-state velocity for $t \to \infty$ is therefore zero. The repositioning time from an end-positioned nucleosome $0N78$ to a nearly completely remodeled state ($37N41$ - 37 bp flanking DNA, i.e., 95% of 39 bp) in 200 s is rather well in accordance with gel mobility shift experiments. Comparing our results with a FRET experiment at 2 μM ATP, we obtain an average velocity of ≈ 0.005 bp/s. This is clearly slower than the experimental value of $\approx 0.05 - 0.1$ bp/s.

Comparing the upper limit of 8/min for the rate k_c' to the 5/min for the initial translocation steps with rates k_1 and k_2, one predicts that the ACF motor will fall off before complete remodeling, as experimentally observed in the earlier experiments. The more recent experiments point to a highly processive motor, i.e., a motor capable of finishing the complete remodeling before falling off. Lowering the rate k_c' to 1/min, which improves the processivity of the motor, increases at 2 μM ATP, the average velocity from ≈ 0.005 bp/s to ≈ 0.05 bp/s, hence by one order of magnitude. The repositioning time decreases from $\approx$ 200 s to $\approx$ 100 s in the case of 10 μM ATP. Consequently, lowering the value of the off-rate k_c' to 1/min, improves the accord with experiments. Figure 4.17 shows the effect of a variation of the ATP concentration on the velocity and diffusion profiles.

The effective diffusion constant in Figures 4.16b and 4.17b is calculated from

$$D_{eff} = \left[\sum_{n=1}^{7} P(n,t)x_n^2 - \Big(\sum_{n=1}^{7} P(n,t)x_n\Big)^2\right]/(2t) \tag{4.22}$$

and reaches a maximal value of 1–2 bp^2/s. We can introduce the Péclet number

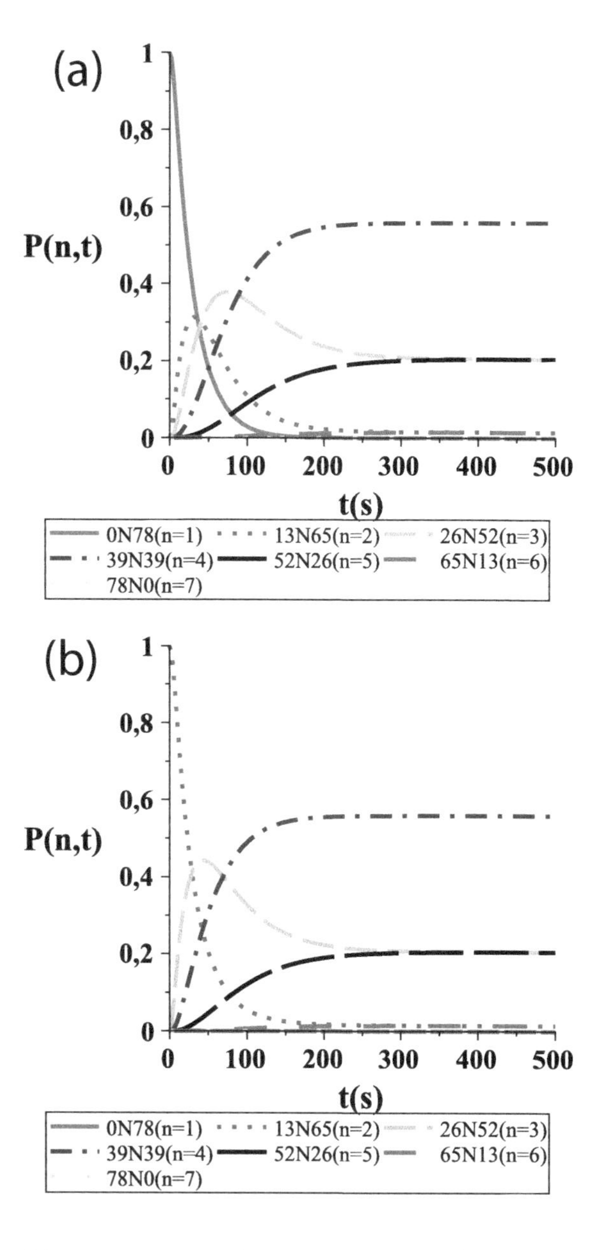

FIGURE 4.15 The probability distribution $P(n, t)$ of the effective one-motor model to find the mononucleosome in a repositioned state n after time t, with the initial conditions $0N78$ (a) and $13N65$ (b). The ATP concentration is in both cases equal to 10 μM ATP (Vandecan and Blossey, 2012). Reprinted with permission from the American Physical Society.

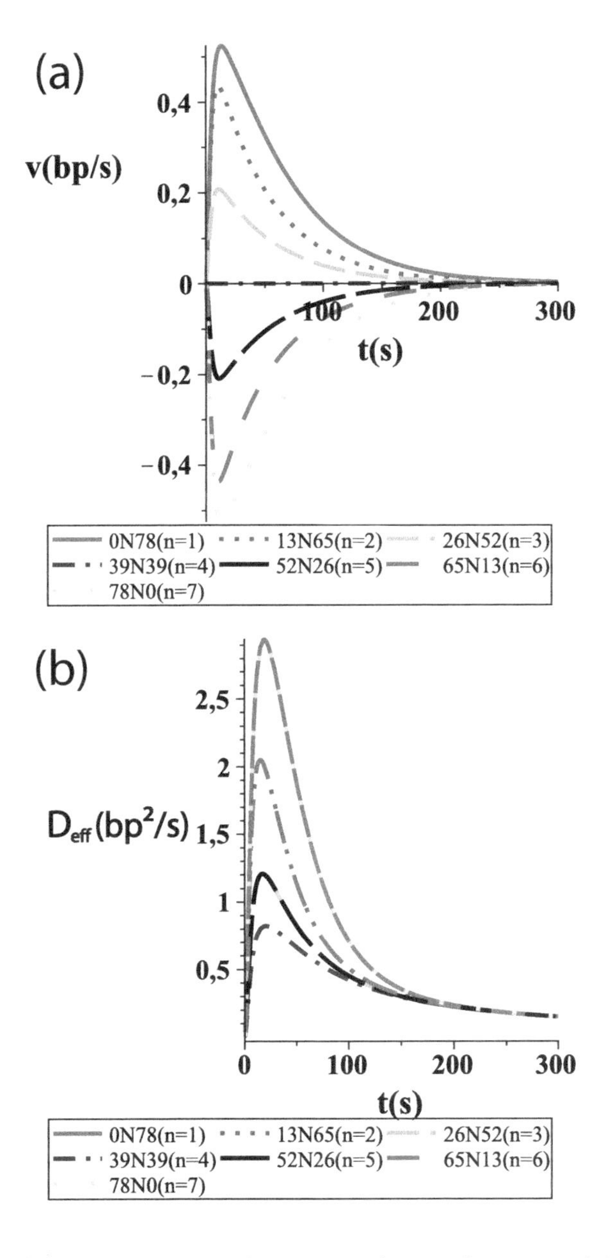

FIGURE 4.16 (a) The motor velocity, or the nucleosome sliding velocity, for the effective one-motor model. The nucleosome sliding velocity reaches a maximum at $t \approx 10$ s to decrease to zero velocity, when the nucleosome is center-repositioned with respect to the DNA. (b) The effective diffusion constant for the different initial positions. The ATP concentration is in both cases equal to 10 μM ATP (Vandecan and Blossey, 2012). Reprinted with permission from the American Physical Society.

as a measure of linear transport compared to diffusion,

$$Pe = \frac{v\ell}{D_{eff}}, \tag{4.23}$$

where v is the velocity, as defined in Eq. (4.21), and ℓ a characteristic length-scale of the system. In our model the Péclet number reaches zero for the time $t \to \infty$, indicating that the velocity profiles decrease more rapidly to zero than the effective diffusion constant. Large position fluctuations can then be expected in steady state, which are indeed observed in the experimental FRET studies due to the bidirectional and processive movement of ACF.

We now turn to the two-motor model. As before, we first look at the probability of finding an initially end-positioned nucleosome over time. The result is shown in Figure 4.18. We assume the same binding and unbinding rate of ATP and ADP, i.e., $k_u = k_p$. Concerning the translocation steps, we let the 13 bp intermediates in our model correspond to the sum of the kinetic steps of 7 bp, 3 bp and again 3 bp (between kinetic pauses), experimentally observed in the initial remodeling of the $0N78$ ($n = 1$) nucleosome. The step time τ_{tr} per intermediate is consequently a simple summation of the individual, ATP-dependent kinetic step times, and $k_1 = 1/\tau_{tr}$. The parameter k_0, independent of the DNA flanking length, can then be determined for all intermediates.

For the kinetic pauses we assume an effective pause τ_k, representing the ATP-loading step, as a sum of the experimentally observed first, second and third kinetic pause. In the motor state (ADP,0) of the intermediate $13N65$, we expect a 50% chance of the direct transition (ADP,0) $\to$ (ATP,0) and a 50% chance of making the transition of (ADP,0) $\to$ (0,ATP), followed by the sampling step (0,ATP) $\to$ (ATP,0). This yields the expression

$$\frac{3}{2}\frac{1}{k_p} = \tau_k, \tag{4.24}$$

which links the effective pause τ_k to the internal motor state transitions k_p and k_u. As a consequence, we have the k_p and k_u for all the intermediates. The values of k_u, k_p and k_0 for several ATP concentrations, as well as the values of the on-rate k_c and the activation step k_I, which are both ATP dependent, are given in Table 4.3.

The probability density functions of the two-motor model are fairly similar to those obtained from the effective one-motor model at the same ATP concentration, although a broader distribution of intermediates is obtained from the two-motor model in steady state. From the probability density distributions, the velocity profiles and the effective diffusion constant as a function of the ATP concentration can again be calculated and are shown in Figure 4.19. The maximal velocity increases from 0.1 bp/s to 5 bp/s as a function of increasing

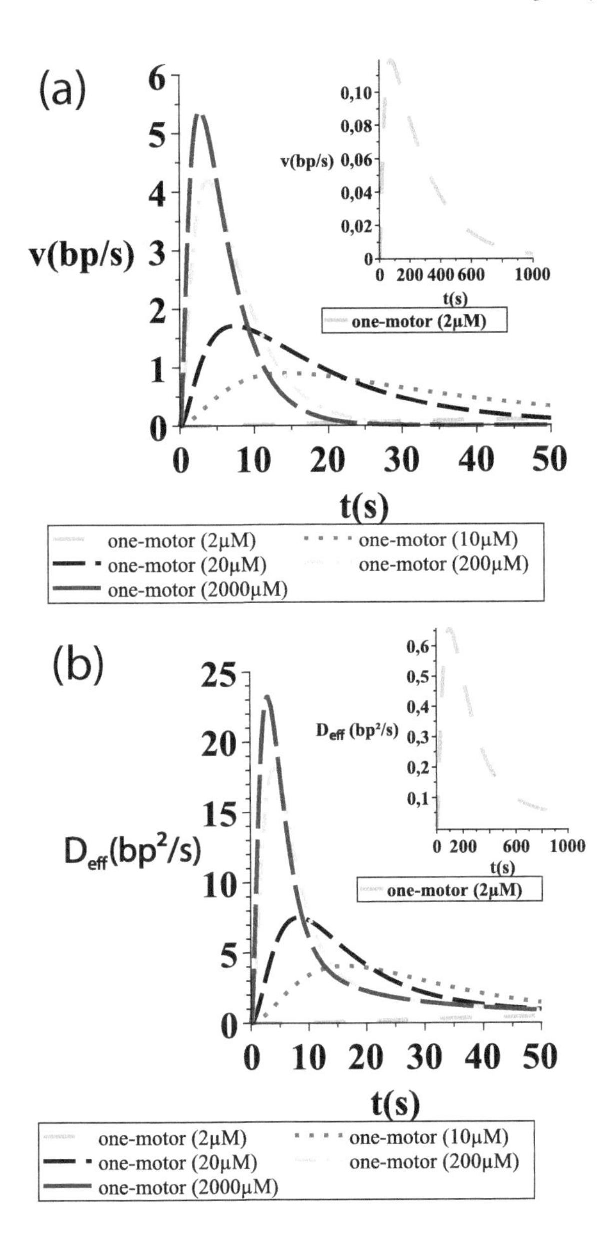

FIGURE 4.17 (a) The motor velocity, or nucleosome sliding velocity, for the effective one-motor model with an initially end-positioned $0N78$ nucleosome (and $k'_c = 1/\text{min}$) at different ATP concentrations. The nucleosome sliding velocity reaches a maximum at $t \approx 10$ s to decrease to zero velocity, when the nucleosome is center-repositioned with respect to the DNA. (b) The corresponding effective diffusion constants for the one-motor model at different ATP concentrations (Vandecan and Blossey, 2012). Reprinted with permission from the American Physical Society.

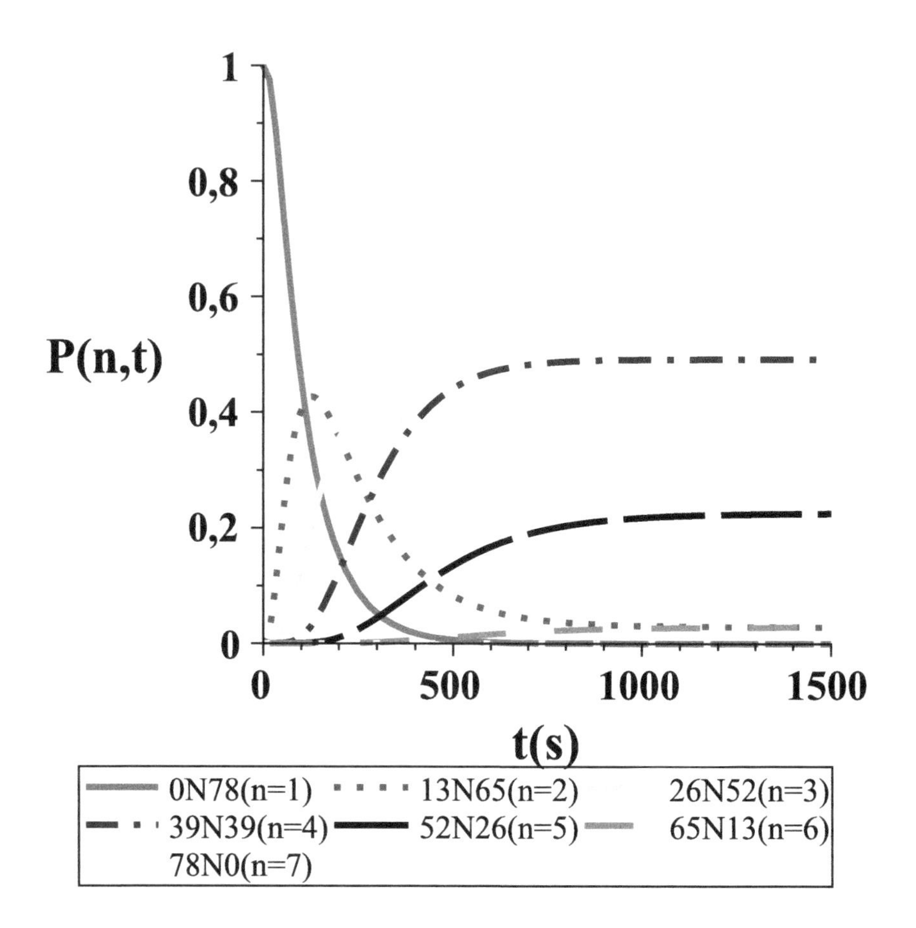

FIGURE 4.18 The probability density function of the intermediates as a function of time, computed from the advanced two-motor model, and thus starting from an initially end-positioned $0N78$ nucleosome. The ATP concentration corresponds to 2 μM (Vandecan and Blossey, 2012). Reprinted with permission from the American Physical Society.

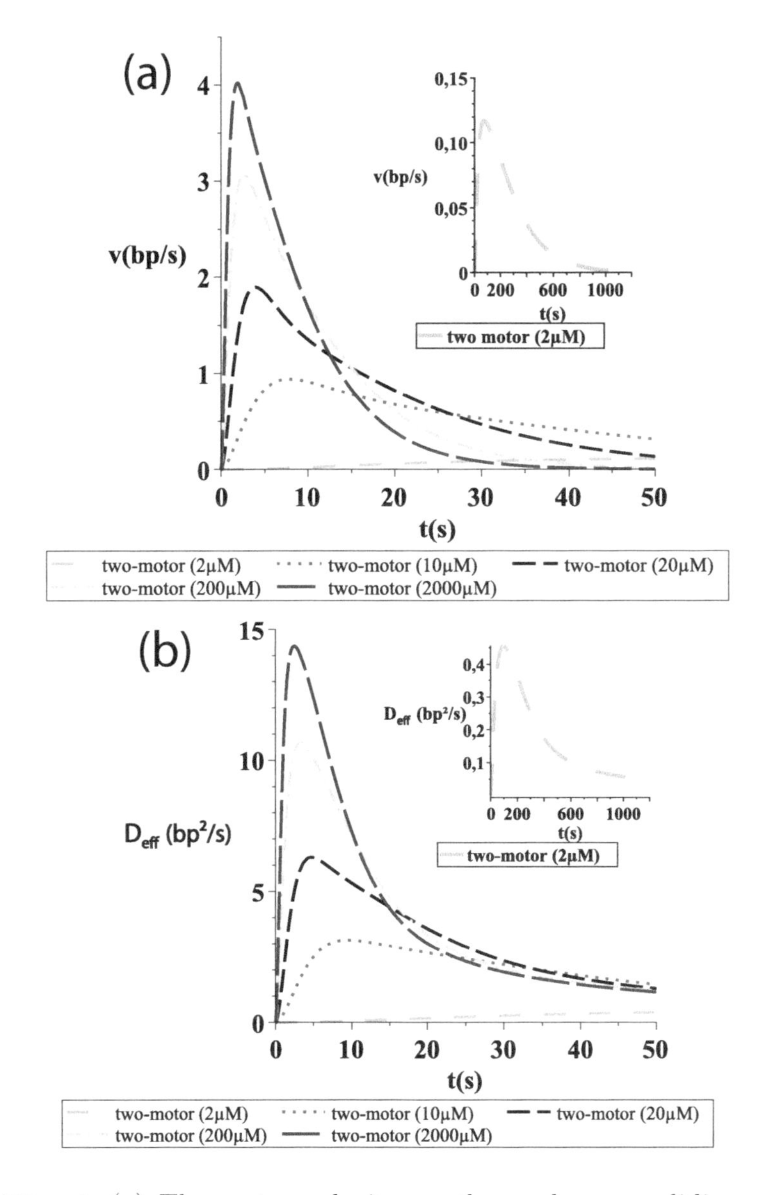

FIGURE 4.19 (a) The motor velocity, or the nucleosome sliding velocity, for the advanced two-motor model with an initially end-positioned $0N78$ nucleosome at different ATP concentrations. The nucleosome sliding velocity reaches a maximum at $t \approx$10–20 s to decrease to zero velocity, when the nucleosome is center-repositioned with respect to the DNA. (b) The corresponding effective diffusion constants for the two-motor model at different ATP concentrations (Vandecan and Blossey, 2012). Reprinted with permission from the American Physical Society.

TABLE 4.3 The estimated values of the model parameters k_0 and $k_p (= k_u)$ as a function of the ATP concentration. The sample values of $1/t_{wait}$ for 20, 200 and 2000 μM are estimated from experiment and we set $k_c = \frac{3}{2}(1/t_{wait})$ and $k_I = 3(1/t_{wait})$. The latter values are multiplied by a factor of three (in a rough approximation) to obtain a system saturated in ACF. The waiting time t_{wait} for 2 μM is taken from the experimentally observed distributions, which has an average of $\approx 50\,$s for a (high) ACF concentration of 16 nM. The values of k_0 and k_p were likewise estimated from the experiments (Vandecan and Blossey, 2012). Reprinted with permission from the American Physical Society.

[ATP] (μM)	k_0 (min^{-1})	k_p (s^{-1})	k_c (s^{-1})	k_I (s^{-1})
2	0.02	0.02	0.03	0.06
10	0.15	0.1	0.2	0.45
20	0.3	0.2	0.45	0.9
200	0.5	0.4	0.7	1.4
2000	0.6	0.6	1.0	2.1

ATP concentration, from 2 μM to 2 mM. While the one-motor model uses effective remodeling rates from normalized Cy3 intensities in FRET studies, and the two-motor model includes experimental kinetic pauses and translocation steps (thus analyzing the FRET time traces in more detail), these experimental results independently give rise to qualitatively similar velocity profiles and effective diffusion constants. We may therefore conclude that our assumptions in the two-motor model, describing a more detailed communication scheme between the two motors and explaining the phenomenon of kinetic pauses as ATP-loading periods, are reasonable.

Both models include an ATP-dependent binding step of ACF, which renders the description more realistic. At high ATP concentrations, the velocities of the effective one-motor model are larger than those of the two-motor model. At low ATP concentrations (up to 20 μM ATP), these differences are negligible. The same applies to the effective diffusion constants for the two models.

Another interesting point to consider is the effect of a change of the rate k_u, which controls the unbinding of ATP from motor 1 and binding of ATP to motor 2, reflecting the sampling between the two motors. Statistically, there is a 50% chance that the ATP molecule is bound to the wrong motor in view of centering the nucleosome with respect to DNA. However, the ATP hydrolysis rate toward the short flanking DNA (e.g., transition rate $k_2' \approx 0.1/\text{min}$) is low so that the ATP molecule unbinds and binds to the ATPase unit of the other motor with a high probability. The ATP hydrolysis rate of the motor toward the longer flanking DNA (e.g., the transition rate $k_1 \approx 5/\text{min}$) is fast so that it is probable that the translocation step is actually done. Putting $k_u = 0$ or setting it to a very low value results in nucleosome–remodeler complexes which are not center-positioned with highest probability in steady state; see Figure 4.20. Thus, this model emphasizes the issue that, after arriving in an ADP state, not only the next ATP loading to the ATPase unit of either motor (the arrow with transition rate k_p) is important, but a subsequent sampling by binding and unbinding of ATP molecules between the ATPase units of the motors is also necessary.

More detailed calculations of the on-rate k_c and of the activation step rate k_I reveal further interesting insights on the processivity of the ACF motor and the robustness of both models. According to Michealis–Menten kinetics of ACF binding followed by an ATP-dependent activation step, the inverse of the waiting time $1/t_{wait}$ is given by

$$1/t_{wait} = \alpha(ATP)\frac{[ACF]}{[ACF] + K_{ACF}}\frac{[ATP]}{[ATP] + K_{ATP}}. \tag{4.25}$$

After curve fitting of experimental data with this formula, we obtain $K_{ACF} \approx 4.8\,\mu$M and $K_{ATP} \approx 117\,\mu$M. The proportionality constant $\alpha \approx 0.7\ s^{-1}$ for high concentrations of [ATP] = 200 and 2000 μM. For low ATP concentrations $\alpha \approx 0.56$ with [ATP] = 20 μM and $\alpha \approx 1.3$ with [ATP] = 2 μM. These minor deviations in α are due to the fact that K_{ACF} is ATP-dependent too, but we lack the experimental data for the ATP dependence of K_{ACF}.

The activation step rate $k_I = 1/\langle t_{lag}\rangle$ with the average lag time $\langle t_{lag}\rangle$ and the binding rate $k_c = 1/\langle t_{bind}\rangle$ with the average binding time $\langle t_{bind}\rangle$ are calculated from the waiting-time distributions of the experimental systems for [ACF] = 4 nM and [ATP] = 2 μM, 20 μM. The lag time is only ATP dependent and the values for high ATP concentrations are predicted by

$$1/\langle t_{lag}\rangle = A\frac{[ATP]}{[ATP] + K_{ATP}}, \tag{4.26}$$

with a fitted value of $A \approx 0.7\,s^{-1}$. Because the waiting time is the sum of the binding and the lag time, i.e., $t_{wait} = t_{lag} + t_{bind}$, Eq. (4.25) permits the computation of the $k_c = 1/\langle t_{bind}\rangle$. All values are gathered in Table 4.4.

For the effective one-motor model, these more detailed k_c and k_I approximations show once again that the k'_c upper limit of 8/min is too high, and point to a highly processive motor with k'_c values even lower than 1/min. A high processivity supports the two-motor model which does not include the dissociation of the remodeler while actively translocating. The velocity profiles with the more rigorous calculations of k_c and k_I are qualitatively similar to profiles generated with the rough approximations of k_c and k_I, but quantitatively tend to lower values.

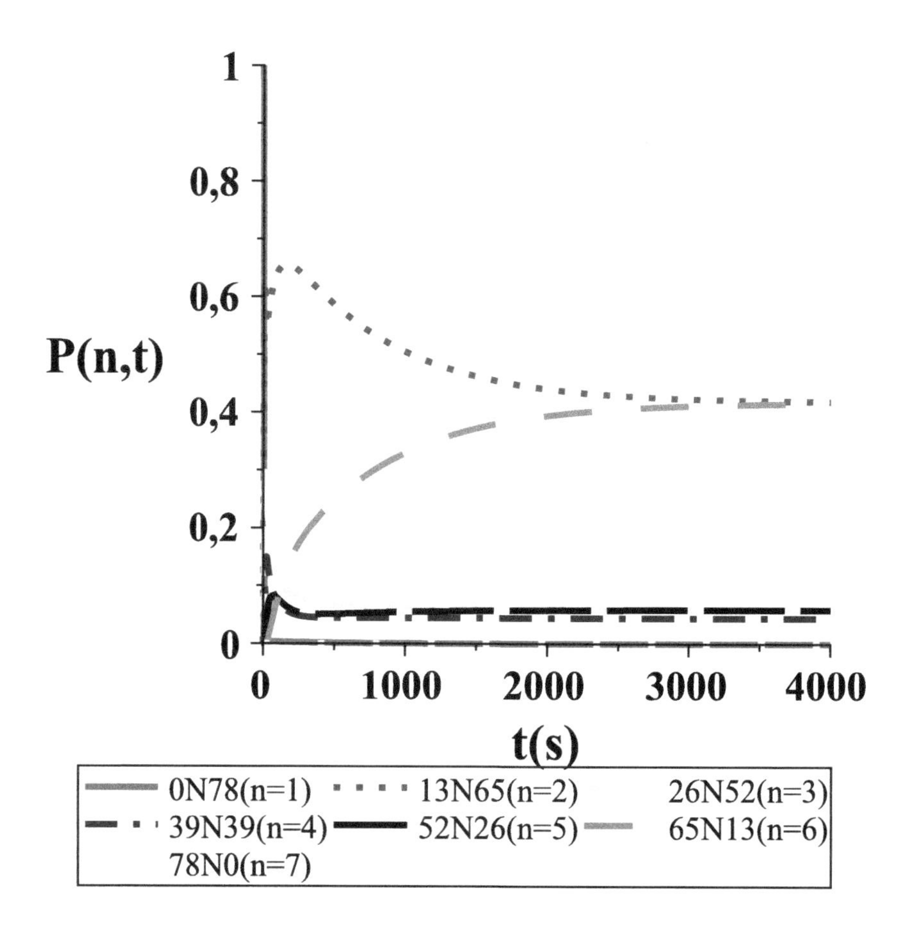

FIGURE 4.20 When the sampling between the two motors is turned off (the value of k_u is reduced by a factor of 100), the probability density function of the advanced two-motor model does not produce center-positioned nucleosomes with highest probability in steady state. The ATP concentration corresponds to 20 μM (Vandecan and Blossey, 2012). Reprinted with permission from the American Physical Society.

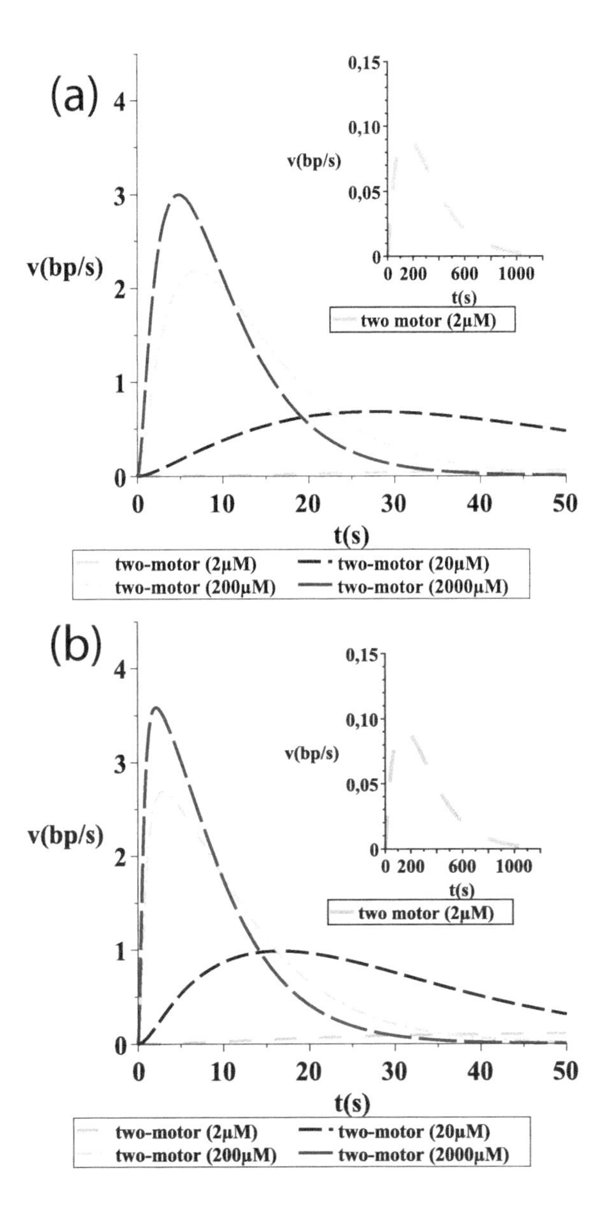

FIGURE 4.21 The motor velocity, or the nucleosome sliding velocity, for the advanced two-motor model with an initially end-positioned $0N78$ nucleosome at different ATP concentrations, using Table 4.4 with [ACF] = 4 nM (a) and [ACF] = 30 nM (b) (Vandecan and Blossey, 2012). Reprinted with permission from the American Physical Society.

TABLE 4.4 The model parameters $k_c = 1/\langle t_{bind}\rangle$ and $k_I = 1/\langle t_{lag}\rangle$ as a function of the ATP-concentration, with [ACF] = 4 nM and a saturated system of [ACF] = 30 nM, from more rigorous approximations (Vandecan and Blossey, 2012). Reprinted with permission from the American Physical Society.

[ATP](μM)	k_c (s^{-1}) [ACF] = 4 nM	k_c (s^{-1}) [ACF] = 30 nM	k_I (s^{-1})
2	0.024	0.1	0.023
20	0.06	0.26	0.1
200	0.36	2.7	0.44
2000	0.55	3.9	0.66

A Fokker–Planck model for nucleosome remodeling by ACF

Although the model described in the previous section is capable of describing many features of the repositioning of nucleosomes by the ACF remodeler, it lacks a key physical feature: the explicit action of the force. In the previous section we modeled the action of the two remodelers with a master equation model which captures the positions assumed by the nucleosome (or, rather, the nucleosome–remodeler complex), with the important ingredient of the remodeler synchronization — but the action of the remodelers is not explicitly present in this description. In this section we introduce a Fokker–Planck description of the same system which does just that. The discussion follows Vandecan and Blossey (2013). While the previous description of the master equation model was very explicit, in this case we limit ourselves to the essentials of this model and the results obtained from it.

In the Fokker–Planck model, we retain the description of the two motors as discussed in the previous section, but we now require the knowledge of the potentials belonging to the chemical states. As these quantities are not available from experiment, we need to model them. These effective driving potentials relate to the consumption of ATP and the resulting power stroke in which the chemical energy is converted into mechanical energy needed to pull on the DNA strand. The pulling of the DNA from the flanking length into the nucleosome leads to a bulge of length $\Delta\ell$ which corresponds to a potential energy decrease ΔU_d. Based on the experimental findings, the slope of the power stroke potential $\Delta U_d/\Delta\ell$ equals a generated force $F \approx k_BT/\text{bp}$.

The chemical states not involved in the conversion of chemical to mechanical energy are mainly governed by the thermal motion of DNA, and will be incorporated into the thermal diffusion constant D_{NR}, which can be due to twist or loop diffusion (see Chapter 1). We estimate it as $D_{NR} = (k_BT/F)v$ using the Einstein relation, where v is the experimental local translocation speed, $v = 13\text{bp}/k_{tr}$, where $1/k_{tr}$ is the effective translocation time needed to move 13 bp along the DNA strand.

The chemical states of the model are easily stated. We denote by state **1** the configuration (0,0), i.e., both remodelers are unloaded. States **2** and **3** are the loaded states (ATP,0) and (0,ATP), respectively. Finally, the two states with ATP burnt to ADP*Pi are **4** and **5**, (ADP*Pi,0) and (0, ADP*Pi). As before, the rates for the irreversible transitions **2** → **4** and **3** → **5** are length dependent. The transition matrix **M** reads as

$$\mathbf{M} = \begin{pmatrix} -2k_p & k_{p,s} & k_{p,s} & k_{tr} & k_{tr} \\ k_p & -k_w(\ell) - k_{p,s} & 0 & 0 & 0 \\ k_p & 0 & -k'_w(\ell) - k_{p,s} & 0 & 0 \\ 0 & k_w(\ell) & 0 & -k_{tr} & 0 \\ 0 & 0 & k'_w(\ell) & 0 & -k_{tr} \end{pmatrix}. \tag{4.27}$$

The dynamics can now be translated into five coupled Fokker–Planck equations that allow us to compute the probability distributions $P_\sigma(\ell, t)$ of the five chemical states, $\sigma = 1 ,\cdots, 5$, as a function of ℓ and time t. They can be written as

$$\partial_t P_\sigma(\ell, t) + \partial_\ell J_\sigma(\ell, t) = \sum_{\sigma'=1}^{5} M_{\sigma\sigma'} P_{\sigma'}(\ell, t), \tag{4.28}$$

with the probability flux

$$J_\sigma(\ell, t) = -\frac{1}{\gamma} \partial_\ell U_\sigma(\ell) P_\sigma(\ell, t) - D_{NR} \partial_\ell P_\sigma(\ell, t) \,. \tag{4.29}$$

Here $U_\sigma(\ell)$ is the potential (as a function of ℓ) in the chemical state σ, D_{NR} denotes the thermal diffusion constant of the nucleosome-remodeler complex, and $\gamma = k_B T/D_{NR}$ is the friction coefficient. The model parameters are listed in Table 4.5.

TABLE 4.5 Estimated values of the model parameters as a function of the ATP concentration (Vandecan and Blossey, 2013). Reprinted with permission from the American Physical Society.

[ATP] (μM)	k_0 (min^{-1})	k_{tr} (s^{-1})	k_p (s^{-1})	D_{NR} (bp^2/s)
2	0.02	0.03	0.03	0.4
20	0.3	0.5	0.3	6.5
200	0.5	0.8	0.6	11
2000	0.6	1	0.9	13

As a first result from the numerical solution of the Fokker–Planck equation, we show the velocity profiles of the nucleosome for different ATP concentrations. The initial distribution for the probability distribution is chosen as a Gaussian

$$P_t(\ell, t = 0) = \frac{1}{2\pi\sigma^2} exp\left(-\frac{(\ell - 10)^2}{2\sigma^2}\right) \tag{4.30}$$

reflecting a family of nearly end-positioned nucleosomes around a mean $\ell = 10$ bp, with a standard deviation $\sigma = 1$ bp. The average velocity profiles $< v >$ are obtained by numerical integration from the expression

$$< v > = \frac{d}{dt} \int_{\ell_i}^{\ell_f} d\ell \, P_t(\ell, t)\ell, \tag{4.31}$$

with ℓ_i and ℓ_f the value of the initial and the final ℓ state, respectively, and

$$P_t(\ell, t) = \sum_{\sigma=1}^{5} P_\sigma(\ell, t) \tag{4.32}$$

is the total probability distribution. Figure 4.22a displays the effect of a change of ATP of four orders of magnitude. Figure 4.22b compares the results that can be obtained from the previous discrete model and the continuum model for two different initial conditions, showing a nice consistency between the two models.

Figure 4.23 shows the near steady-state probability distributions of the Fokker-Planck equation as a function of the sampling rate. As seen before in the discrete two-motor model, a reduction in the sampling rate $k_{p,s}$ leads to off-center positioned nucleosomes.

Positioning multiple nucleosomes with ACF

ACF, like other ISWI remodelers, is relevant for the controlled positioning of nucleosomes along DNA in order to repress genes. It is thus immediately interesting to ask what role this positioning effect has for an array of nucleosomes. This effect can easily be studied with *kinetic Monte Carlo simulations* of remodeler positioning by including a linker-length-dependent positioning rate. The kinetic Monte Carlo technique is highlighted in the grey box.

Figure 4.24 shows the results from such simulations, first reproducing the experimental data (Racki et al., 2009). Kinetic Monte Carlo is nothing but solving the master equation, so in a way we are just repeating what we did before by way of simulations. Indeed, the parametrization of the model is not crucially different from what we discussed before so that we skip these details here. The model parametrization that allowed us to reproduce the centering experiments in stochastic simulations can then also be used for the study of

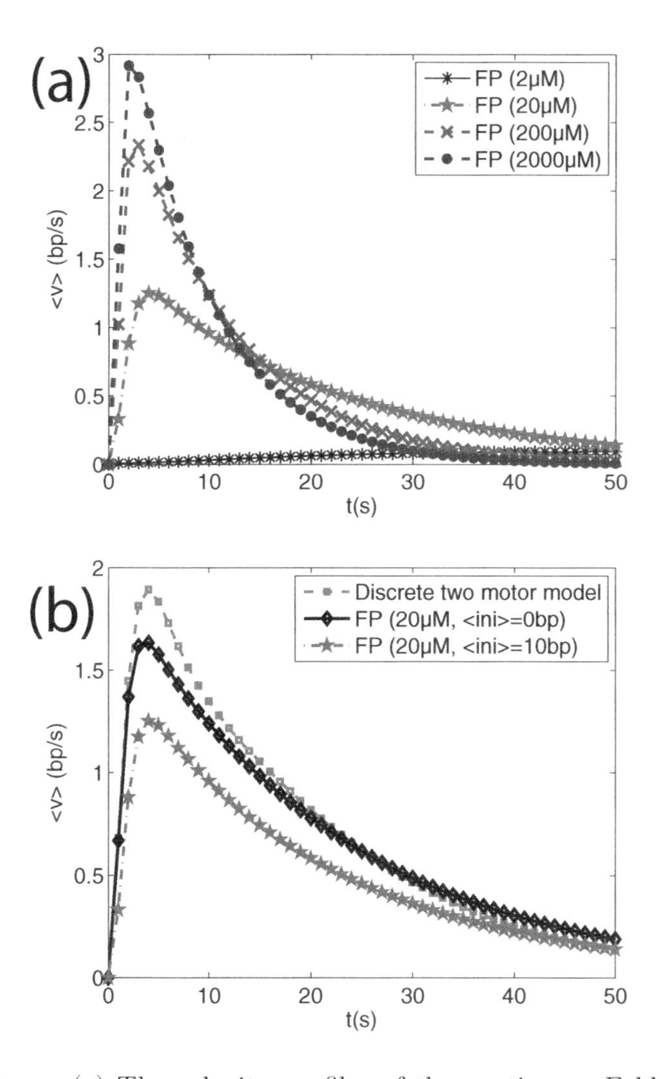

FIGURE 4.22 (a) The velocity profiles of the continuum Fokker–Planck model for different concentrations of ATP, starting from a Gaussian initial distribution with mean $< ini > = 10$ bp and standard deviation $\sigma = 1$ bp. (b) A comparison of the velocity profiles of the discrete two-motor model with the Fokker–Planck model for two distinct initial distributions with mean $< ini >= 0, 10$ bp and standard deviation $\sigma = 1$ bp (Vandecan and Blossey, 2013). Reprinted with permission from the American Physical Society.

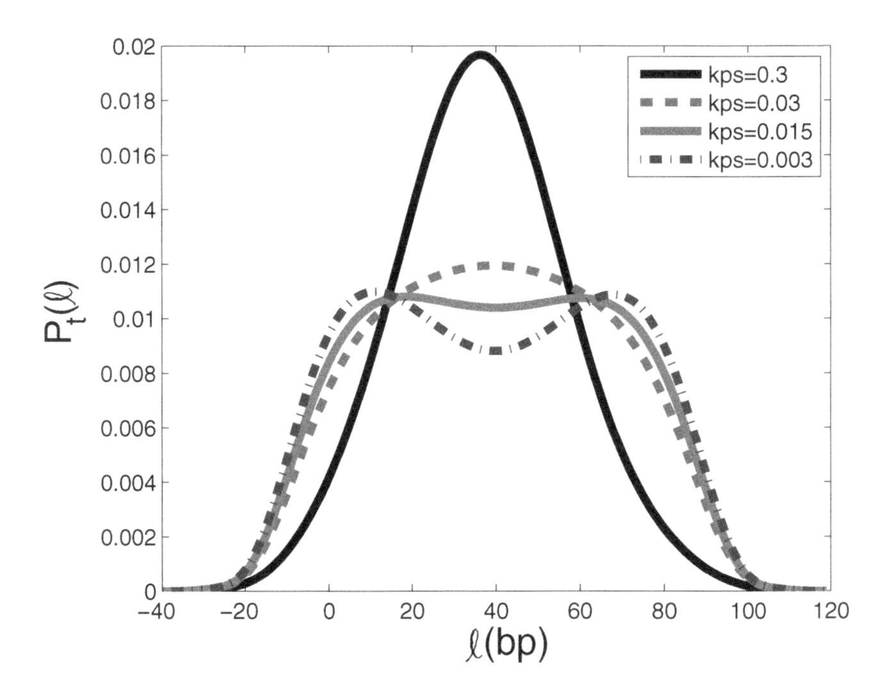

FIGURE 4.23 Near steady-state probabilities of the continuum Fokker–Planck model as a function of a reduced sampling rate $k_{p,s}$ (with an ATP concentration of 20 μM). The time t equals $50, 300, 400$ and 500 s for the diminished sampling rates $k_{p,s} = 0.3, 0.03, 0.015$ and 0.003 (in units of s^{-1}), in that order. At a reduction of ≈ 0.05 times the initial value, the highest steady-state probability does not belong to the center-positioned ℓ state anymore. The distribution turns from a monomodal to a bimodal behavior (Vandecan and Blossey, 2013). Reprinted with permission from the American Physical Society.

arrays, and the results are in shown in Figure, 4.25. The result is also compared to the Kornberg–Stryer behavior we discussed in Chapter 1. The effect on positioning due to a linker-length-dependent positioning rate is clearly visible in this graph: the nucleosomes become ordered over much longer scales away from the strongly positioned nucleosome than in the Kornberg–Stryer case.

Kinetic Monte Carlo. Kinetic Monte Carlo algorithms can be used to model the dynamics of biochemical reactions. In computational biology they often run under the name of the Gillespie algorithm, which is one version of a kinetic MC. Put in a different way, KMC can also be seen as a way to solve the master equation of the system under study with simulation means. A presentation of the Gillespie algorithm can, e.g., be found in Blossey (2006).

Taking all these combined hints from genome-wide analysis of nucleosome positions, both in vitro and in vivo, and the results from single-molecule assays on nucleosome remodeling together, they can finally be combined with genome-wide remodeling experiments which compare wild-type to mutant behaviors of chromatin remodelers. Micrococcal nuclease digestion experiments in *S. cerevisiae* thereby revealed the organization of nucleosomes around transcription start sites in wild-type and gene-deletion mutants, in which ISWI variants and another remodeler, Chd1, were absent (Gkikopoulos et al., 2011). The experiments demonstrate the importance of these remodelers for the specific positioning near the transcription start site; this is particularly the case for the remodeler Chd1. The effect is strongest for highly transcribed genes, but it is also perceptible in low-level transcribed genes. When the data are averaged over all genes, the effect is completely washed out. This averaging procedure is clearly fairly brutal, as even slight differences in nucleosome positions over all genes will lead to cancellations in the data.

The picture controlling the positioning of nucleosomes now becomes more clear. It is a concerted effect of DNA sequence, which can steer preferences for nucleosomes, together with the collective effect due to the presence of multiple nucleosomes, yielding to a basic statistical positioning effect; and ultimately it has an active component based on chromatin remodelers. The decisive "strongly positioned nucleosome" barrier requires the active effect of remodelers. The relative role of these effects has been studied recently in some detail with kinetic Monte Carlo simulations (Parmar, Marko and Padinthateeri, 2014). Such simulations have the enormous advantage that they can switch on and off these different contributions at will, something which is much more difficult of course in experimentation.

If we now understand better how the different contributions act in concert, one element still is not satisfactorily linked in. As we saw in the beginning of

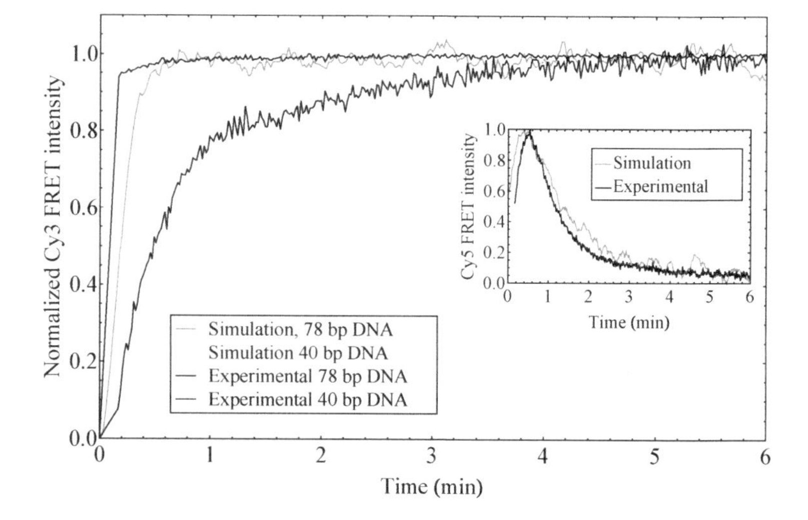

FIGURE 4.24 Model fitting to normalized FRET intensity data from nucleosome positioning experiments on positioning sequences of different base composition and length, for a DNA end-positioned and a DNA interior-positioned fluorophore (inset). 40 bp and 78 bp refer to the length of the linker DNA. The fit to experiment is performed by varying the still unspecified step size of the motor; an optimal result is obtained for a step size of 13 bp (Florescu, Schiessel and Blossey, 2012). Reprinted with permission from the American Physical Society.

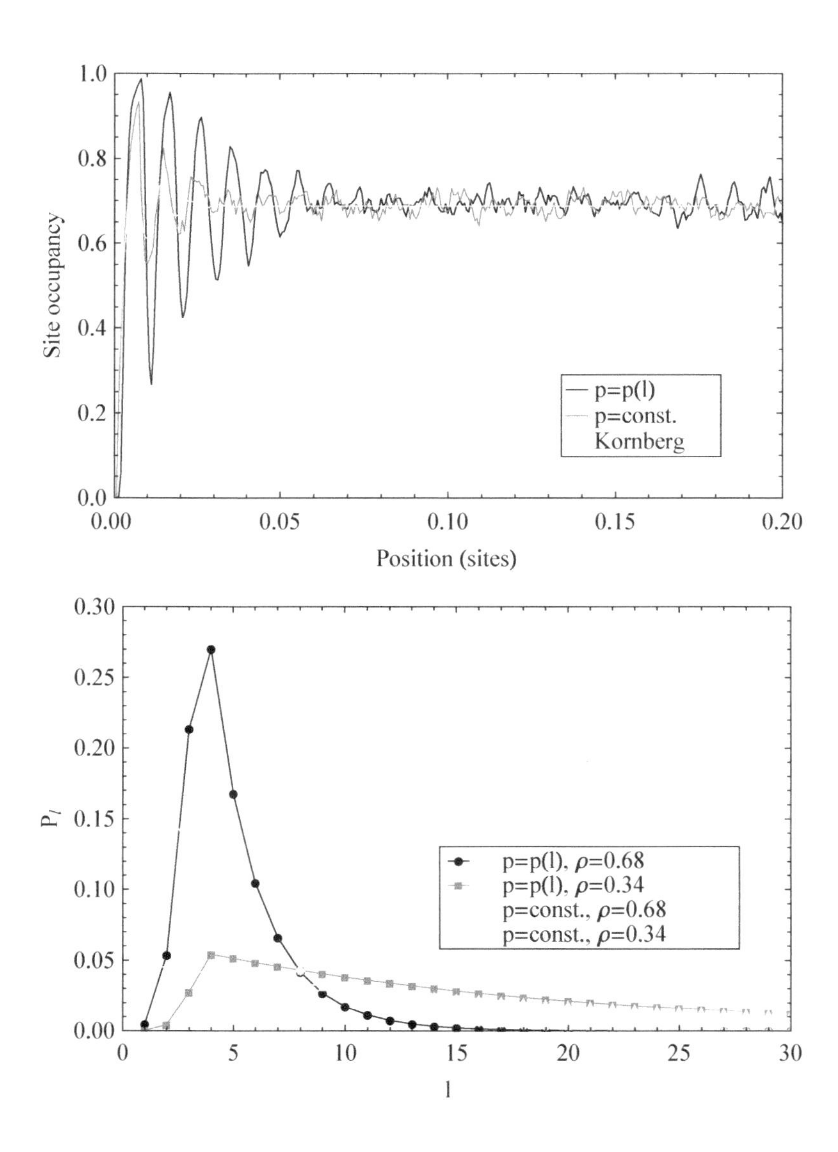

FIGURE 4.25 Top: Site occupancy of a nucleosomal array of 1600 sites with 100 nucleosomes, density $\varrho = 0.68$, after a simulation run of $t = 60$ min. Bottom: Distribution of linker lengths P_ℓ at two densities and for both length-dependent and constant rates. All data were taken after individual runs for 60 min and sampling over 100 realizations (Florescu, Schiessel and Blossey, 2012). Reprinted with permission from the American Physical Society.

this chapter, the analysis of just a few histone modifications alone could also reproduce positioning patterns of nucleosomes. How can this be explained? This question turns out to be linked to another question: how do the remodelers "know" which nucleosome to remodel? We will look for an answer in the next chapter.

References

Allfrey V.G, R. Faulkner and A.E. Mirsky
Acetylation and methylation of histones and their possible role in the regulation of RNA synthesis.
Proc. Natl. Acad. Sci. 51, 786-794 (1964)

Blosser T.R., J.G. Yang, M.D. Stone, G.J. Narlikar and X. Zhuang
Dynamics of nucleosome remodelling by individual ACF complexes.
Nature 462, 1022-1028 (2009)

Blossey R.
Computational biology from a statistical physics perspective.
Chapman & Hall (2006)

Chaban Y., C. Ezeokonkwo, W.H. Chung, F. Zhang, R.D. Kornberg, B. Maier-Davis, Y. Lorch and F.J. Asturias
Structure of a RSC-nucleosome complex and insights into chromatin remodeling.
Nat. Struct. & Mol. Biol. 15, 1272-1277 (2008)

Clapier C.R. and B.R. Cairns
The biology of chromatin remodeling complexes.
Annu. Rev. Biochem. 78, 273-304 (2009)

Filippakopoulos P., S. Picaud, M. Mangos, T. Keates, J.-P. Lambert, D. Barsyte-Lovejoy, I. Felletar, R. Volkmer, S. Müller, T. Pawson, A.C. Gingras, C.H. Arrowsmith and S. Knapp
Histone recognition and large-scale structural analysis of the human bromodomain family.
Cell 149, 214-231 (2012)

Filippakopoulos P. and S. Knapp
Targeting bromodomains: epigenetic readers of lysine acteylation.
Nat. Rev. Drug Discovery 13, 337-355 (2014)

Florescu A.-M., H. Schiessel and R. Blossey
Kinetic control of nucleosome displacement by ISWI/ACF chromatin remodelers.
Phys. Rev. Lett. 109, 118103 (2012)

Gkikopoulos T, P. Schofield, V. Singh, M. Pinskaya, J. Mellor, M. Smolle, J.L. Workman, G.J. Barton, T. Owen-Hughes
A role for Snf2-related nucleosome-spacing enzymes in genome-wide nucleosome organization.
Science 333, 1758-1760 (2011)

Hudson B.P., M.A. Martinez-Yamout, H.J. Dyson and P.E. Wright
Solution structure and acetyl-lysine binding activity of the GCN5 bromodomain.
J. Mol. Biol. 304, 355-370 (2000)

Karlić R., H.-R. Chung, J. Lasserre, K. Vlahoviček and M. Vingron
Histone modification levels are predictive for gene expression.
Proc. Natl. Acad. Sci. (USA) 107, 2926-2931 (2010)

Leschziner A.E., A. Saha, J. Wittmeyer, Y. Zhang, C. Bustamante, B.R. Cairns and E. Nogales
Conformational flexibility in the chromatin remodeler RSC observed by electron microscopy and the orthogonal tilt reconstruction technique.
Proc. Natl. Acad. Sci. (USA) 104, 4913-4918 (2007)

Marmorstein R. and M.-M. Zhou
Writers and readers of histone acetylation: structure, mechanism and inhibition.
Cold Spring Harb Perspect Biol 6:a018762 (2014)

Mühlbacher F., H. Schiessel and C. Holm
Tail-induced attraction between nucleosome core particles.
Phys. Rev. E 74, 031919 (2006)

Nguyen V.Q., A. Ranjan, F. Stengel, D. Wei, R. Aebersold, C. Wu and A.E. Leschziner

Molecular architecture of the ATP-dependent chromatin-remodeling complex SWR1.
Cell 154, 1220-1231 (2013)

Owen D.J., P. Ornaghi, J.C. Yang, N. Lowe, P.R. Evans, P. Ballario, D. Neuhaus, P. Filetici, A.A. Travers AA
The structural basis for the recognition of acetylated histone H4 by the bromodomain of histone acetyltransferase Gcn5p.
EMBO J. 19, 6141-6149 (2000)

Parmar J.J., J.F. Marko and R. Padinhateeri
Nucleosome positioning and kinetics near transcription-start-site barriers are controlled by interplay between active remodeling and DNA sequence.
Nucl. Acids Res. 42, 128-136 (2014)

Polach K.J., P.T. Lowery and J. Widom
Effects of core histone tail domains on the equilibrium constants for dynamic DNA site accessibility in nucleosomes.
J. Mol. Biol. 298, 211-223 (2000)

Racki L.R., J.G. Yang, N. Naber, P.D. Partensky, A. Acevedo, T.J. Purcell, R. Cooke, Y. Cheng and G.J. Narlikar
The chromatin remodeller ACF acts as a dimeric motor to space nucleosomes.
Nature 462, 1016-1021 (2009)

Rippe, K., A. Schrader, P. Riede, R. Strohner, E. Lehmann and G. Längst
DNA sequence- and conformation-directed positioning of nucleosomes by chromatin-remodeling complexes.
Proc. Natl. Acad. Sci. (USA) 104, 15635-15640 (2007)

Singh R.P., G. Brysbaert, M. F. Lensink, F. Cleri and R. Blossey
Kinetic proofreading of chromatin remodeling: from gene activation to gene repression and back.
AIMS Biophysics 2, 398-411 (2015)

Sirinakis G., C.R. Clapier, Y. Gao, R. Viswanathan, B.R. Cairns and Y. Zhang
The RSC chromatin remodelling ATPase translocates DNA with high force and small step size.
EMBO J. 30, 2364-2372 (2011)

Strahl B.D. and C.D. Allis

The language of covalent histone modifications.
Nature 403, 41-45 (2000)

Vandecan Y. and R. Blossey
Stochastic description of single nucleosome repositioning by ACF remodelers.
Phys. Rev. E 85, 061920 (2012)

Vandecan Y. and R. Blossey
Fokker-Plank description of single nucleosome repositioning by dimeric chromatin remodelers.
Phys. Rev. E 88, 012728 (2013)

Yang J.G., T.S. Madrid, E. Sevastopoulos and G.J. Narlikar
The chromatin-remodeling enzyme ACF is an ATP-dependent DNA length sensor that regulates nucleosome spacing.
Nat. Struct. Mol. Biol. 13, 1078-1083 (2006)

Further reading

Andrews F.H., B.D. Strahl and T.G. Kutateladze
Insights into newly discovered marks and readers of epigenetic information.
Nat. Chem. Biol. 12, 662-668 (2016)

Bannister A.J. and T. Kouzaridis
Regulation of chromatin by histone modifications.
Cell Res. 21, 381-395 (2011)

Hopfner K.-P., C.-B. Gerhold, K. Lakomek and P. Wollmann
Swi2/Snf2 remodelers: hybrid views on hybrid molecular machines.
Curr. Op. Struct. Biol. 22, 225-233 (2012)

Lavelle C., E. Praly, D. Bensimon, E. Le Cam and V. Croquette
Nucleosome-remodelling machines and other molecular motors observed at the single-molecule level.
FEBS Journal 278, 3596-3607 (2011)

Perner J., J. Lasserre, S. Kinkley, M. Vingron and H.-R. Chung
Inference of interactions between chromatin modifiers and histone modifications: from ChIP-Seq data to chromatin-signaling.

Nucl. Acids Res. 42, 13689-13695 (2014)

Rothbart S.B. and B.D. Strahl
Interpreting the language of histone and DNA modifications.
Biochim. Biophys. Acta 1839, 627-643 (2014)

Schreiber S.L. and B.E. Bernstein
Signaling network model of chromatin.
Cell 111, 771-778 (2002)

An alternative view of the role of post-translational modifications of histone tails.

Turner B.M.
Chromatin and gene regulation.
Blackwell Science, London (2001)

This is a well-written introduction to the field despite being somewhat older.

CHAPTER 5

Regulating chromatin

Regulating chromatin remodeling

Since the remodelers carry recognition domains for histone tails, the assumption that at least certain modifications, like acetylation, are linked to the action of the remodeling motors is an obvious one to make. Indeed, the possible relation between these two aspects appears in the literature in many places, perhaps most explicitly in the perspective article by Cosgrove, Boeke and Wolberger (2004) published in *Nature Structural & Molecular Biology* with the title "Regulated nucleosome mobility and the histone code". The schematic drawing presented in this paper and shown here as Figure 5.1 couples the action of histone acetylases (HATs) and histone deacteylases (HDACs) to the activities of remodelers, but how this works precisely is not made very clear in this figure. In the following we will develop a quantitative approach to this scenario.

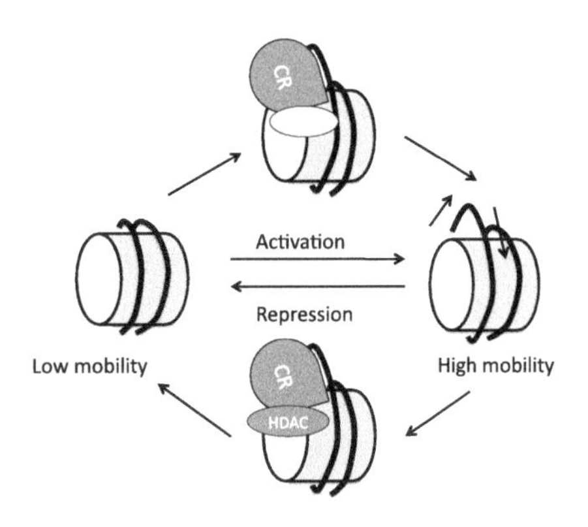

FIGURE 5.1 The regulated nucleosome mobility model, adapted from Cosgrove, Boeke and Wolberger (2004). CR refers to ATP-dependent chromatin remodelers, HAT: histone acetylase, HDAC: histone deactylase.

As a first step we will look into the kinetics of remodelers under the effect of modifications to the histone tails. Ferreira, Flaus and Owen-Hughes (2007) studied the effect of acetylations of the histone tails on the enzymatic activity of chromatin remodelers. We can remember at this point the finding discussed in Chapter 1 that the histone tails did not dramatically affect the intrinsic mobility of the nucleosomes. Ferreira et al. determined two kinetic constants for the remodeling reactions. In order to understand these, we look little bit further into the mathematics of enzyme reaction kinetics, following the classic work of Michaelis and Menten.

Consider the reaction

$$R + N \rightleftharpoons_{k_-}^{k_+} RN \rightarrow_{k_{cat}} N^* + R \tag{5.1}$$

in which N and R are nucleosome and remodeler, respectively. The ratio k_-/k_+ defines the dissociation constant K_d, and the reaction speed (not the remodeler speed) is then given by

$$v = k_{cat}[RN] \tag{5.2}$$

where $[RN]$ is the concentration of the nucleosome–remodeler complexes. Following Michaelis–Menten, this concentration is given by

$$[RN] = \frac{[R_{total}] \cdot [N]}{K_M + [N]} \tag{5.3}$$

where $[R_{total}]$ is the concentration of remodelers both bound and free, and K_M is the Michaelis–Menten constant,

$$K_M = \frac{k_{cat} + k_-}{k_+} . \tag{5.4}$$

We can write the reaction speed as

$$v = \frac{k_{cat} \cdot [R_{total}] \cdot [N]}{K_M + [N]} \tag{5.5}$$

wherein $k_{cat} \cdot [R_{total}]$ is the maximal reaction speed v_{max}, which is reached at high nucleosome concentrations, $[N] \gg K_M$. Thus we have

$$v = v_{max} \frac{[N]}{K_M + [N]} . \tag{5.6}$$

Putting $K_M = [N]$, we find that at this value, $v = v_{max}/2$, which allows us to interpret K_M as the constant where half of the enzymes are engaged.

The hyperbolic Michaelis–Menten law was measured by Ferreira et al. for differently acetylated histone tails under the action of different remodelers. We discuss the results for two cases, RSC and ISW2, which we encountered in the previous chapter. For RSC, it was found that the catalytic constant k_{cat} is hardly different between wild-type nucleosomes and various combinations of acetylations on H3, H4 and both. The Michaelis–Menten constant, however, reduces by about a factor of three for acetylations involving the H3 tail, or combined tetra-acetylations of both H3 and H4 tails.

ISW2 behaves very differently. When compared with wild type nucleosomes, acetylation of H4 mildly increases K_M and mildly reduces k_{cat}. A truncation of the tail also has no significant affect on K_M. However, it quite strongly, by about a factor of 5, reduces the catalytic rate k_{cat}.

These findings provide clear hints that histone modifications in fact play a significant role in the regulation of the enzyme kinetics of chromatin remodelers. We now turn to a scenario which can rationalize these findings.

Blossey and Schiessel (2008) proposed that the existing link between histone tail modifications and chromatin remodelers allows for a *kinetic proofreading scheme*, as we have already seen before in the context of toposiomerases. Here we will now get a little deeper into matters.

Kinetic proofreading was first proposed by Hopfield (1974) and Ninio (1975) as a mechanism of error correction — or even better, for error prevention in biochemical processes. The role of kinetic effects in transcriptional regulation were discussed as early as 1993 by Herschlag and Johnson (1993). A particularly lucid introduction to this mechanism can be found in Uri Alon's book (Alon, 2007), which explains it in its original context, the regulation of the specificity of mRNA translation —the experimental evidence was published in (Hopfield et al., 1976). It is useful to go through this case first. In this example the mechanistic action of the remodeler is absent; here, one is concerned with enzymatic actions only.

Following Alon, let us denote by C the codon on an mRNA in the ribosome that is going to be translated into a protein. (A codon is a three-letter word of bases.) This process is controlled by the binding of the correct tRNA, which carries the corresponding amino acid, called c. In an equilibrium binding reaction, we would have

$$c + C \rightleftharpoons^{k'_c}_{k_c} (cC) \rightarrow_v \text{correct amino acid placed} \tag{5.7}$$

and we have $K_c = k'_c/k_c$ as the dissociation constant, and the incorporation rate $R_{corr} = v[c] \cdot [C]/K_c$. We can likewise imagine a competing reaction in which a wrong amino acid would be inserted. This would then happen along the reaction scheme

$$d + C \rightleftharpoons^{k'_d}_{k_d} (dC) \rightarrow_v \text{incorrect amino acid placed} \tag{5.8}$$

which gives rise to a dissociation constant K_d, and an incorporation rate of $R_{incorr} = v[d] \cdot [C]/K_d$. We would obviously have $R_{incorr} < R_{corr}$ and their ratio

$$F_0 \equiv \frac{R_{incorr}}{R_{corr}} = \frac{K_c}{K_d} \approx \frac{k'_c}{k'_d} \tag{5.9}$$

is a measure of the specificity of the amino acid insertion reaction. Note that, again, we have assumed that the specificity is in the dissociation reaction.

Even if the dissociation constants between the presumed correct or incorrectly placed amino acid were to differ by a factor of 100, this cannot explain the extremely low error in mRNA translation, as it must be.

Now consider the following scheme:

$$\begin{array}{l} c + C \rightleftharpoons^{k'_c}_{k_c} (cC) \rightarrow_m (c^*C) \rightarrow_v \text{correct amino acid placed} \\ \qquad\qquad\qquad\quad \downarrow_{\ell'_c} \\ \qquad\qquad\qquad\quad c + C \end{array} \tag{5.10}$$

In this reaction, after the binding reaction there is an irreversible reaction with rate m which modifies c into c^*. Further, the modified reactant (c^*C) can fall off with rate l'_c, but not directly get back into the reaction. Alon invokes the image of a one-way revolving door in an exhibition for the effect of this reaction. In this scenario, the relative concentration of the modified and non-modified reaction products is given by

$$[c^*C] = \frac{m}{l'_c}[cC] \tag{5.11}$$

and this rate prefactor nows enters into the error rate, which is given by

$$F \equiv \frac{R_{incorr}}{R_{corr}} = \frac{K_c}{K_d}\frac{l'_c}{l'_d} \approx \left(\frac{k'_c}{k'_d}\right)^2 = F_0^2\,, \tag{5.12}$$

where we have assumed that the dissociation rate of the modified and non-modified reaction products are fairly equal.

We can now immediately transfer the insights gained from this reasoning to the case of chromatin remodeling, building on the two experimental facts: first, chromatin remodelers "read" the histone tail state refers to the binding of a remodeler recognition domain with a rate k', and an unbinding with a rate k. We take the rate of unbinding as specific: the rate of binding k' is determined by the frequency of molecular collisions, while the stability of the binding between the remodeler recognition domain and the histone tail decides on the rate of unbinding k. The second fact is that the active engagement of the remodeler in displacing the nucleosome is irreversible and consumes ATP; irreversible thus means that in order to undo the induced motion, additional

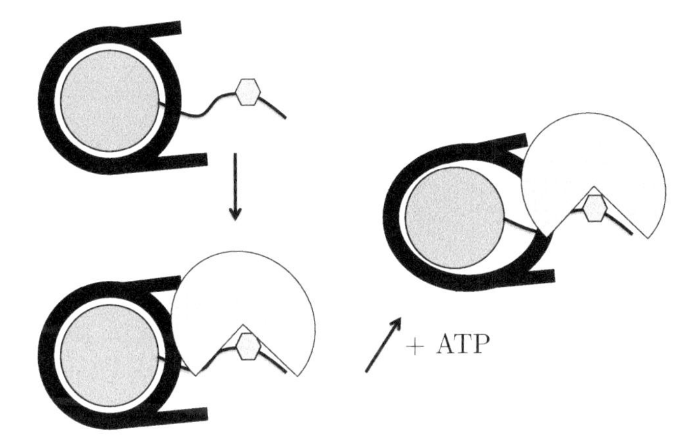

FIGURE 5.2 Schematic view of the remodeling reaction as proposed by Blossey and Schiessel (2008). Top left: a histone tail modification is present to help recruit the remodeler shown bottom left. Right: the recruited remodeling complex recognizes the modification and lifts the DNA from the histone octamer under ATP consumption.

ATP consumption is required. This step equals the modification step of the transfer tRNA discussed before.

If we denote the remodeler by R, the nucleosome by N and the remodeler–nucleosome complex by I, the activated complex by I^* and the mobile complex by M, the proofreading scenario is given by the following reaction scheme in which m' is the ATP-dependent activation rate, t the translocation rate of the nucleosome–remodeler complex, and ℓ' the dissociation rate of the activated remodeler complex from the nucleosome (Blossey and Schiessel, 2011):

$$\begin{array}{c} R + N \rightleftharpoons_{k}^{k'} I \rightarrow_{m'} I^* \rightarrow_{t} M \\ \downarrow_{\ell'} \\ R + N \end{array} \tag{5.13}$$

The reaction scheme is also depicted schematically in Figure 5.2. The irreversible step occurs repeatedly if the remodeler acts in a processive manner.

We note that, of course, the rate scheme we adopt here is a very simplified one. In Figure 5.2, the binding of a transcription factor can also been added; we will turn to this aspect shortly. But first we turn the above reaction equations into rate equations for concentrations,

$$\frac{d[I]}{dt} = k'[R][N] - (m' + k)[I], \tag{5.14}$$

$$\frac{d[I]^*}{dt} = m'[I] - (\ell' + t)[I^*]. \tag{5.15}$$

From these equations the ratio

$$f \equiv \frac{[I^*]}{[R][N]} \tag{5.16}$$

can be calculated when stationarity of the reaction is assumed, in this way characterizing its discriminatory capacity. One finds

$$f \equiv \frac{m'k'}{(\ell + t)(m' + k)}. \tag{5.17}$$

As before, we now compare two cases which differ in the presence or absence of histone tail modifications on the nucleosomes. The reactions are described by ratios f_1 and f_2, whose ratio in turn gives the error fraction spelled out as

$$F \equiv \frac{m_1'k_1'}{m_2'k_2'} \frac{(k_2 + m_2')}{(k_1 + m_1')} \frac{(\ell_2' + t_2)}{(\ell_1' + t_1)} \tag{5.18}$$

We will now elaborate on these relations in two particular contexts, one concerning gene activation, the other gene repression.

Problem 6. Write down rate equations for the kinetic proofreading scenario shown in Figure 2.17 for the action of topoisomerase II.

Two examples: gene activation and repression

The IFN-β gene

In the first example — for gene activation — we address the case of the IFN-β gene, which is involved in immune response (Ford and Thanos, 2010). Genes can quite generally be grouped into three classes: those that are constitutively active, the so-called "housekeeping" genes (we encountered those before in nucleosome positioning); those that are constitutively switched off, as they fulfill no specific task any more in a differentiated cell; and finally those, particularly interesting ones, that need to be shut down in a healthy cell, but

a)

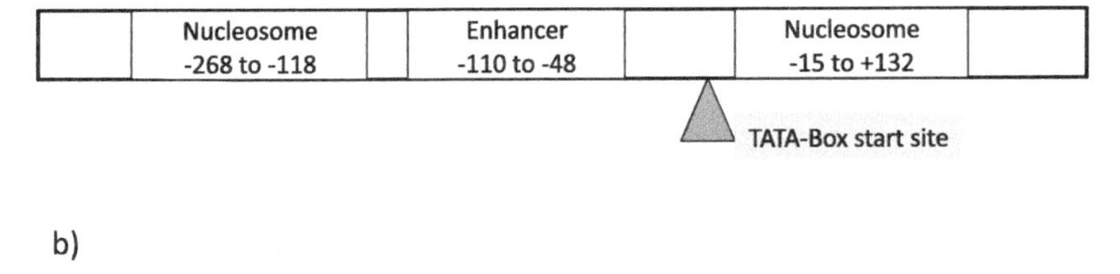

b)

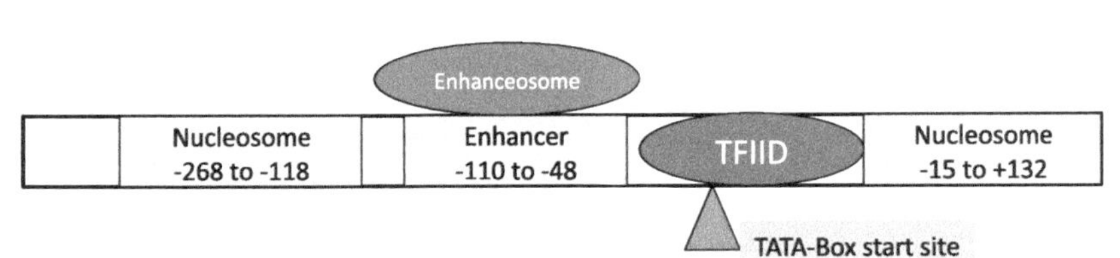

FIGURE 5.3 The regulatory region of the IFN-β gene. (a) An enhancer element is located upstream of the gene in a nucleosome-free region (NFR). The TATA-box of the gene is partially occluded by a nucleosome. (b) After remodeling, the nucleosome is moved downstream, allowing the general transcription factor TFIID to bind. Drawn after Agalioti et al. (2000, 2002). The numbers in both graphs denote the position relative to the IFN-β transcription start site on the DNA sequence *before* remodeling, as the nucleosome has no precise position afterwards (Singh et al., 2015).

rapidly activated upon, e.g., a viral infection. The IFN-β gene is an example for such a case.

Figure 5.3 shows a schematic representation of the build-up of the IFN-β gene and its regulatory regions. The first key element is an *enhancer region*, which is positioned in a nucleosome free region (NFR) in between two positioned nucleosomes (Figure 5.3a). The downstream nucleosome partially occludes the TATA-box transcription factor binding site upstream of the gene. In order to pre-initiate transcription of this gene, the downstream nucleosome must be remodeled in order to shift it further downstream to allow for the generic transcription factor TFIID to bind.

How can the IFN-β gene go from the inactive state shown in Figure 5.3a, to the transcription initiation-ready state shown in Figure 5.3b? The first step is the assembly of an enhanceosome which consists of a transcription factor complex with a set of three different transcription factors engaged in triggering the inflammation response following a viral infection, among them the well-known NF-κB factor (Agalioti et al., 2000). The enhanceosome then recruits a histone-tail modification writer, GCN5 (which we encountered before in Chapter 4), which in turn modifies the histone tails of the nucleosome. On histone H4, lysine residue K8 is acetylated while on H3, K9 and K14 are acetylated (Agalioti et al., 2002). Acetylation of H4-K8 allows the recruitment of a remodeling complex, SWI-SNF, which recognizes the modification via its bromodomains. The remodeler then shifts the nucleosome downstream. Acetylation of H3-K4/K9 recruits the general transcription factor TFIID and hence the IFN-β gene is ready for transcription.

While Agalioti et al. established the time series of these events, they did not quantify the process by their corresponding free energies of binding or dissociation. At present it is therefore not useful to cast the process into rate equations, which would in fact correspond to a rather large system of equations, as given above. Qualitatively, we can distinguish between three successive steps:

i) The build-up of the *enhanceosome* is analogous to transcription factor recruitment in prokaryotes; see our discussion in Chapter 3. For the IFN-β gene, the free energies controlling the build-up of the transcription factor complex could in principle be studied in a similar manner, at least in vitro.

ii) The intermediate step is the recruitment of the histone tail writer GCN5, which occurs via protein–protein interactions with the enhanceosome, and its action on the histone tails. We have quantified the specificity of this reaction before (see Chapter 4).

iii) The remodeler SWI-SNF acts on the nucleosome covering the TATA box next to the IFN-β gene.

A difficulty with this example is that the different steps involve so many protein–protein and protein–DNA interactions that a complete quantification of this process is hard to conceive in practice. The second example, in fact, relies on in vitro data, and constitutes the first quantified case for kinetic proofreading in chromatin remodeling.

ISWI/ACF

In the previous chapter we encountered the remodeler ACF from the ISWI family of remodelers. G.J. Narlikar proposed a kinetic proofreading scenario for this system, and, based on the remodeling assays, could provide the first quantitative estimates for the rates involved (Narlikar, 2010). This allowed her to determine several of the parameters in the Hopfield formula. In the notation used by Narlikar we can rewrite the equation as

$$F \equiv \frac{k_{I,1}k_1'}{k_{I,2}k_2'}\frac{(k_2 + k_{I,2})}{(k_1 + k_{I,1})}\frac{(k_{off,2} + k_{tr_2})}{(k_{off,1} + k_{tr,1})} \tag{5.19}$$

where the following parameters were determined by her work: $k_{I,1} = 20/\text{min}$; $k_{I,2} = 1/\text{min}$; $k_{off,1} = 8/\text{min}$, $k_{off,2} = 160/\text{min}$; $k_{tr,1} = 80/\text{min}$, $k_{tr,2} = 80/\text{min}$. For the remaining parameters we assume $k_1' = k_2'$ and $k_i = \ell_i' = k_{off,i}$. When these numbers are used, an error factor of the correct substrate of $F \approx 313$ is obtained. This estimate is indeed of the same order of magnitude as the one put forward in Blossey and Schiessel (2008). The experiments by Erdel et al. (2010) yield qualitative evidence for the scenario as well.

It is noteworthy that ISWI remodelers are found to have a peculiar mechanism of auto-regulation (Clapier, Nightingale and Becker, 2002; Clapier and Cairns, 2012). ISWI contains an identical "basic patch" sequence motif of histone H4, and thus enters into a competitive interaction with the tail. This effect can also be integrated into the kinetic proofreading scenario (Brysbaert, Lensink and Blossey, 2015).

Epigenetics

In this last section we return to the histone code — this time on a more global, cellular level. We will look at an example of gene regulation which allows the tuning of the cells of an organism to adapt to specific environmental conditions.

We have seen already that histone marks — in our case acetylation or its absence — can be essential in the recruitment of the molecular interaction partners, notably the chromatin remodeling enzymes. Beyond this immediate "mechanical" effect, required to make a gene "readable", other types of signals and their dynamic evolution can code for complex developmental programs.

We will see such an example in the case of *vernalization*, which is the induction of flowering in plants after a period of cold.

In order to get an idea of such a mechanism, we begin, however, with a mathematical model of such a tuning process and follow some very recent work by Jost (2014). The starting point of his model is a DNA region containing n nucleosomes. Each of the nucleosomes is assumed to be in one of three states (obviously, in reality, there can be several more): state U, an unmarked nucleosome state A, an "active" nucleosome, and I, an "inactive" nucleosome. The notions of active and inactive refer to histone tail marks that either favor remodeling (acetylation in the active case) or disfavor it (e.g., by certain methylations in the inactive case). Basically, the model makes use of the three state transitions

$$I \rightleftharpoons U \rightleftharpoons A\,. \tag{5.20}$$

How do the nucleosomes pass from one state to the other? There are two mechanisms assumed, one specific and one non-specific. In the specific mechanism, nucleosomes recruit histone tail modifiers at a given rate; otherwise, the states are supposed to also change randomly with a (lumped) rate k_0. The reaction rates corresponding to the above scheme then read as

$$r_{U\to A} \equiv (k_0 + \varepsilon_A \varrho_A)(n - n_A - n_I)\,, \tag{5.21}$$

$$r_{U\to I} \equiv (k_0 + \varepsilon_I \varrho_I)(n - n_A - n_I)\,, \tag{5.22}$$

$$r_{A\to U} \equiv (k_0 + \varepsilon_I \varrho_I) n_A\,, \tag{5.23}$$

$$r_{I\to U} \equiv (k_0 + \varepsilon_A \varrho_A) n_I\,, \tag{5.24}$$

where the specific rates are written as rate factors times the local density of the modification states.

For a physicist, this looks very much like a three-state Ising or Potts model (spin up, down or neutral), with specific interactions being the spin–spin interaction, and the random transitions corresponding to thermally induced spin flips. In this analogy, one can consider the difference between active an inactive nucleosomes, $m \equiv (n_A - n_I)/n$ (the "magnetization") as the order parameter to describe the dynamics of the system.

Simplifying the rate parameters $\varepsilon_{A,I}$ to a single value ε, one can write down mean-field-type dynamical equations for the densities of the histone marks:

$$\frac{d\varrho_A}{dt} = (k_0 + \varepsilon\varrho_A)(1 - \varrho_A - \varrho_I) - (k_0 + \varepsilon_I)\varrho_A\,, \tag{5.25}$$

$$\frac{d\varrho_I}{dt} = (k_0 + \varepsilon\varrho_I)(1 - \varrho_A - \varrho_I) - (k_0 + \varepsilon_A)\varrho_I\,. \tag{5.26}$$

This system of equations has, depending on parameters, three fixed points, which in terms of m read as

$$m_0 = 0\,,\ m_{\pm} = \pm(k_0/\varepsilon)\sqrt{(\varepsilon/k_0 + 1)(\varepsilon/k_0 - 3)} \tag{5.27}$$

for $\varepsilon > 3k_0$. This is a classic dynamical system which displays a supercritical pitchfork bifurcation at a critical point $\varepsilon_c = 3k_0$. In order to go beyond the mean-field approximation, one can write down a Fokker–Planck equation for the probability $P(m)$,

$$\partial_t P(m) + \partial_m \left([w_+(m) - w_-(m)]P(m) + \frac{1}{2n}\partial_m [w_+(m) - w_-(m)]P(m) \right) = 0 \tag{5.28}$$

with the propensities to increase or decrease m given by

$$w_+ \equiv (r_{U\to A} + r_{I\to U})/n\,,\ w_+ \equiv (r_{U\to I} + r_{A\to U})/n\,. \tag{5.29}$$

This Fokker–Planck equation is a diffusion equation for a particle in a one-dimensional potential. In the case of bistability, the mean first passage time to pass from one state to the other via the neutral state is given by the expression

$$\tau \equiv \frac{18\pi}{(\varepsilon - \varepsilon_c)\sqrt{3(\varepsilon/k_0 + 3)}} \exp[V(0) - V(m)]\,, \tag{5.30}$$

where $V(m) = -\ln P_\infty$ where P_∞ is the steady-state probability distribution. It is notable that τ scales exponentially with n, a factor of relevance for the relaxation of the change from one state to another.

Adding the factor of asymmetry, i.e., to include again the difference between ε_A and ε_I, one obtains the stability diagram shown in Figure 5.4b, in which three regions appear: two distinct monostable and one bistable region. The latter only arises when the recruitment factors are strong, which is the case in a strongly competing situation between two states.

This example shows that under very simple and generic assumptions the epigenetic state of nucleosomes can show bistable behaviour, which is a prerequisite for the establishment of regulatory patterns that can be encoded by specific modifications.

We now turn to a biological example that can be covered by such a mathematical model in a very specific way. The example, as mentioned before, concerns the establishment of epigenetic silencing in the vernalization of plants, more specifically in the model organism *Arabidopsis thaliana*. *Vernalization* is the mechanism by which plants perceive winter and block premature blossoming, i.e., out of season, as might be the case after only a short interval of cold. The perception of the winter period must therefore be capable of duration detection. The situation is sketched in Figure 5.5.

Arabidopsis' genome contains a gene called FLC ("Flowering Locus C") which

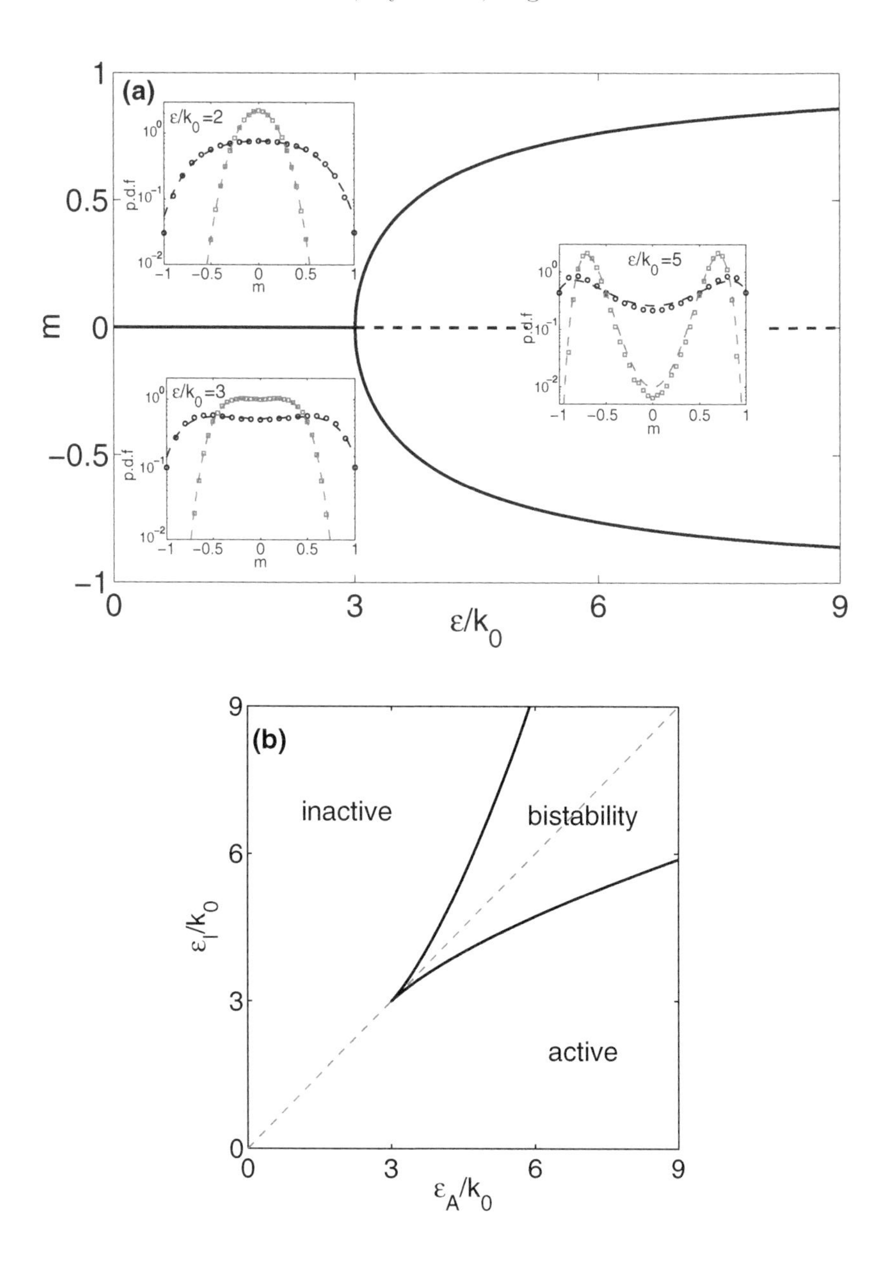

FIGURE 5.4 (a) Nucleosome tail states as a function of tail modification rates in the symmetric regime. The pitchfork bifurcation occurs at the value $\varepsilon = 3k_0$ in the mean-field case (continuous lines); the inset graphs sketch solutions of a Fokker–Planck equation and Gillespie-based simulations of the master equation, based on the rates (5.29). (b) Stability diagram and boundaries for the asymmetric case. Reprinted with permission from Jost (2014). © American Physical Society.

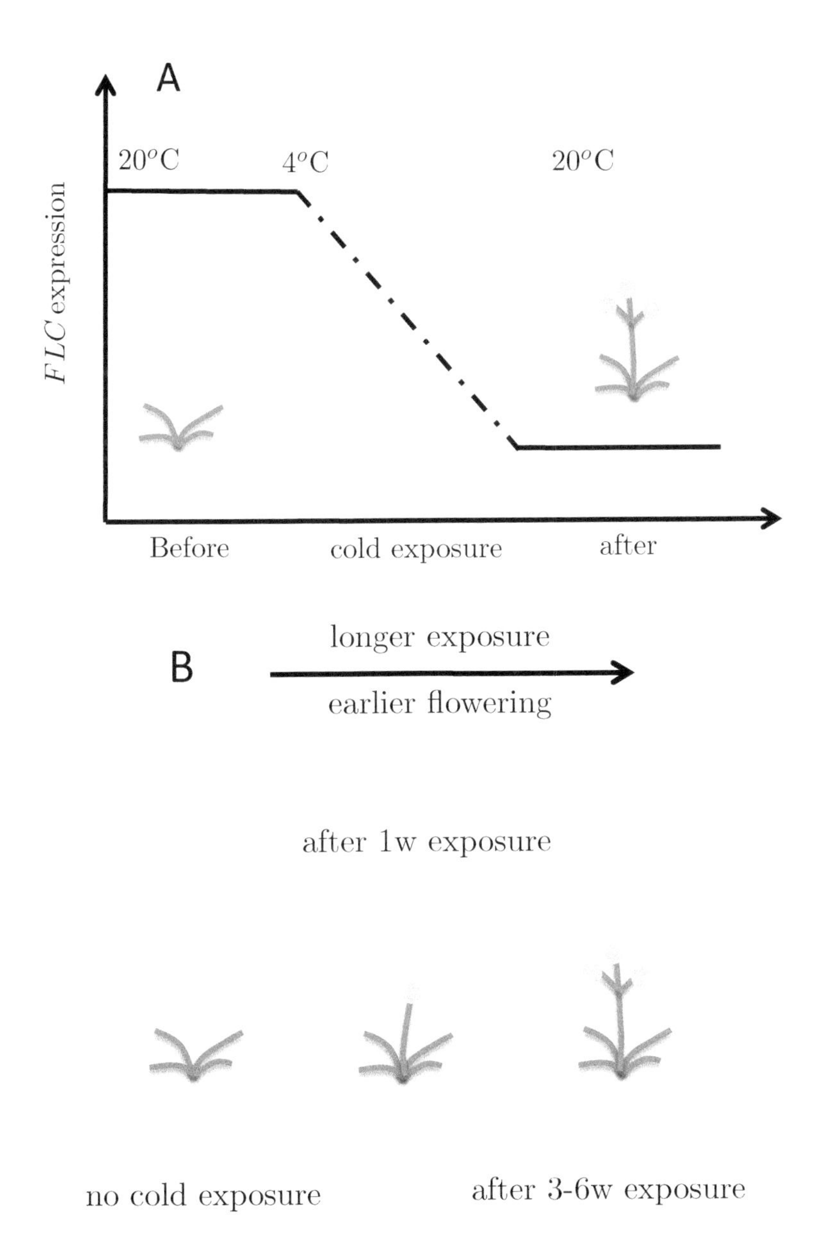

FIGURE 5.5 (A) The floral repressor gene FLC is highly expressed before exposure to cold (summer, autumn). The expression level decreases steadily over the cold exposure period and remains stable at a low level when cold exposure has ended. (B) The expression level attained during cold exposure determines the onset of flowering. Adapted from Song et al. (2012).

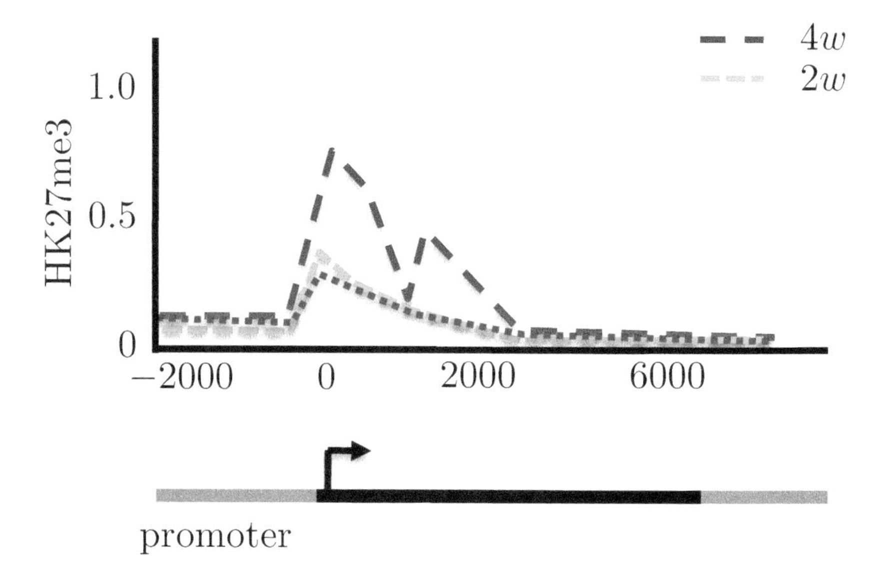

FIGURE 5.6 Qualitative behavior of ChIP data for FLC before and after cold exposure with duration indicated; baseline: no vernalization. Adapted with simplifications from Angel et al. (2011).

represses flowering. This gene is itself repressed by the positioning of trimethylations on the lysine residue 27 of histone H3, a mark which is set by the Polycomb repressive complex PRC2. The exposure to a cold period leads to a repressed state of the gene which is limited to one locus, the immediate vicinity of the FLC gene. The establishment of the gene silencing during the cold period and its maintenance after cold until flowering occurs can be clearly distinguished: the degree of FLC silencing depends on the period of cold exposure. Finally, this epigenetic state is not inherited from one generation of plants to the next.

Qualitatively, what one expects is that in the pre-cold season, FLC is strongly expressed (there are no flowers in summer or autumn), a level which decreases during the cold exposure period. It then remains stably repressed upon the return to warm temperatures. The positioning of the histone mark H3K27me3 during cold exposure occurs in a localized position near exon 1 of the FLC gene. Figure 5.6 sketches ChIP data for FLC, comparing non-vernalized plants and plants after exposure for two and four weeks, respectively. The increase of H3K27me3 levels at four weeks in the so-called "nucleation" region is clearly visible. Upon increase of the cold exposure duration, the histone mark propagates across the gene. Upon the end of cold exposure, the attained level of histone mark H3K27me3 spreads out evenly across the gene.

References

Agalioti T., S. Lomvardas, B. Parekh, J. Yie, T. Maniatis and D. Thanos
Ordered recruitment of chromatin modifying and general transcription factors to the IFN-β promoter.
Cell 103, 667-678 (2000)

Agalioti T., G. Chen and D. Thanos
Deciphering the transcriptional histone acetylation code for a human gene.
Cell 111, 381-392 (2002)

Alon U.
An introduction to systems biology.
Chapman & Hall CRC, London (2007)

Angel A., J. Song, C. Dean and M. Howard
A Polycomb-based switch underlying quantitative epigenetic memory.
Nature 476, 105-109 (2011)

Blossey R. and H. Schiessel
Kinetic proofreading of gene activation by chromatin remodelers.
HFSP Journal 2, 167-170 (2008)

Blossey R. and H. Schiessel
Kinetic proofreading in chromatin remodeling: the case of ISWI/ACF.
Biophys. J. 101, L30-L32 (2011)

Brysbaert G., M.F. Lensink and R. Blossey
Regulatory motifs on ISWI chromatin remodelers: molecular mechanisms and kinetic proofreading.
J. Phys.: Condens. Matter 27, 064108 (2015)

Clapier C.R., K.P. Nightingale and P.B. Becker
A critical epitope for substrate recognition by the nucleosomal remodeling ATPase ISWI.
Nucl. Acids Res. 30, 649-655 (2002)

Clapier C.R. and B.R. Cairns
Regulation of ISWI involves inhibitory modules antagonized by nucleosomal epitopes.

Nature 492, 280-284 (2012)

Cosgrove M.S., J.D. Boeke and C. Wolberger
Regulated nucleosome mobility and the histone code.
Nat. Struct. Mol. Biology 11, 1037-1043 (2004)

Erdel F., T. Schubert, C. Marth, G. Längst and K. Rippe
Human ISWI chromatin-remodeling complexes sample nucleosomes via transient binding reactions and become immobilized at active sites.
Proc. Natl. Acad. Sci. (USA) 107, 19873-19878 (2010)

Ferreira H., A. Flaus and T. Owen-Hughes
Histone modifications influence the action of Snf2 family remodelling enzymes by different mechanisms.
J. Mol. Biol. 374, 563-579 (2007)

Ford E. and D. Thanos
The transcriptional code of human IFN-β gene expression.
Biochim. et Biophys. Acta 1799, 328-336 (2010)

Herschlag D. and F.B. Johnson
Synergism in transcriptional activation: a kinetic view.
Genes & Dev. 7, 173-179 (1993)

Hopfield J.J.
Kinetic proofreading: a new mechanism for reducing errors in biosynthetic processes requiring high specificity.
Proc. Natl. Acad. Sci. 71, 4135-4139 (1974)

Hopfield J.J., T. Yamane, V. Yue and M. Coutts
Direct experimental evidence for kinetic proofreading in amino acylation of tRNA.
Proc. Natl. Acad. Sci. 73, 1164-1168 (1976)

Jost D.
Bifurcation in epigenetics: implications in development, proliferation, and diseases.
Phys. Rev. E 89, 010701(R) (2014)

Narlikar G.J.

A proposal for kinetic proof reading by ISWI family chromatin remodeling motors.
Curr. Op. Chem. Biol. 14, 660-665 (2010)

Ninio J.
Kinetic amplification of enzyme discrimination.
Biochimie 57, 587-595 (1975)

Singh R.P., G. Brysbaert, M. F. Lensink, F. Cleri and R. Blossey
Kinetic proofreading of chromatin remodeling: from gene activation to gene repression and back.
AIMS Biophysics 2, 398-411 (2015)

Song J., A. Angel, M. Howard and C. Dean
Vernalization - a cold-induced epigenetic switch.
J. Cell Sci. 125, 3723-3731 (2012)

Further reading

Cortini R., M. Barbi, B.R. Caré, C. Lavelle, A. Lesne, J. Mozziconacci and J.-M. Victor
The physics of epigenetics.
Rev. Mod. Phys. 88, 025002 (2016)

Erdel F. and K. Rippe
Binding kinetics of human ISWI chromatin-remodelers to DNA repair sites elucidate their target location mechanism.
Nucleus 2, 105-112 (2011)

Erdel F. and K. Rippe
Chromatin remodelling in mammalian cells by ISWI-type complexes - where, when and why?
FEBS Journal 278, 3608-3611 (2011)

Sneppen K. and I.B. Dodd
A simple histone code opens many paths to epigenetics.
PLoS Comp. Bio. 8, e1002643 (2012)

A key paper developing the idea of the histone code into a mathematical ap-

proach.

Swygert S.G. and C.L. Peterson
Chromatin dynamics: interplay between remodeling enzymes and histone modifications.
Biochim. et Biophys. Acta 1839, 728-736 (2014)

There has been recently a renewed interest in kinetic proofreading within the the statistical physics community. Some papers are

Hartich D., A.C. Barato and U. Seifert
Nonequilibrium sensing and its analogy to kinetic proofreading.
New J. Phys. 17, 055026 (2015)

Murugan A., D.A. Huse and S. Leibler
Speed, dissipation, and error in kinetic proofreading.
Proc. Natl. Acad. Sci 109, 12034-12039 (2012)

Murugan A., D.A. Huse and S. Leibler
Discriminatory proofreading regimes in nonequilibrium systems.
Phys. Rev. X 4, 021016 (2014)

Rao R. and L. Peliti
Thermodynamics of accuracy in kinetic proofreading: Dissipation and efficiency trade-offs.
J. Stat. Mech. P06001 (2015)

CHAPTER 6

Chromatin beyond transcription: splicing, repair and replication

In this chapter of the book we will take a look beyond the basic transcription (or, for that matter, repression) of a single gene in the context of chromatin. It will come as no surprise that full genome data will play a key role in this chapter — so far we encountered them in the discussion of nucleosome positioning and the placement of histone marks around transcriptional start sites. In Chapters 2 and 4 we saw how the DNA sequence together with collective effects determine where nucleosomes are placed, and how they are dynamically regulated with the help of histone marks and ATP-dependent remodeling processes. In as much as each gene is regulated by its own set of transcription factors, additional levels of regulation come into play on the level of epigenetic marks. These can act as simple stop and go signs, very much like traffic lights, but they can also provide essential clues to how a gene will be read over time; in the traffic analogy, this is like speed control on a highway. It allows epigenetic programming of the activation and repression of genes.

But then, there never is only one gene to be activated or repressed at a given time. In eukaryotes, the gene itself is a fluid quantity, composed of multiple sequences that are even placed at distinct distances from each other. How is the reading of these coordinated in the chromatin context? This will be our first topic. Subsequent sections will address how DNA is repaired in the chromatin context and how the genome is duplicated by the process of replication.

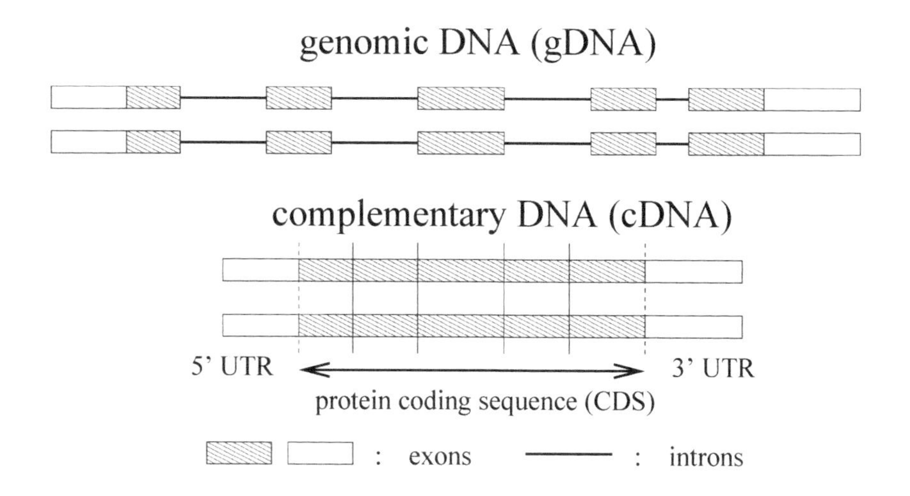

FIGURE 6.1 Definition of exons and introns (Carlon, Malki and Blossey, 2005). Reprinted with permission from the American Physical Society.

Exons and introns

Figure 6.1 conveys a basic idea about exons and introns by looking at genomic DNA and at cDNA, *complementary DNA*. Genomic DNA is generally composed of sequence elements that code for protein domains, the exons, but these portions are not necessarily contiguous in eukaryotes. In fact, they are broken into pieces by the insertion of introns. These need to be removed by a dedicated biochemical machinery, the *spliceosome*. The fact that exons exist gives room for a large variation: proteins that are translated can originate from different exon elements. Complimentary DNA, cDNA, is frequently used in biochemical assays; it corresponds to DNA of a proper gene (exons only), and can be obtained from retranscribed mRNA products after splicing has occurred.

How does one define the exon and introns along DNA? The start position of an intron sequence at its $5'$-end and its ending at the $3'$-end are determined by specific sequence signals that, however, are not entirely conserved. There is also a physical signature to distinguish between introns and exons which is based on the different CG and AT content of the corresponding sequences: exons are richer in CG and are hence more stable. Interestingly, also a difference between different exons can be detected. Looking at the thermal denaturation of cDNA, one can in fact identify exon positions as the whole cDNA sequence opens in a stepwise manner. Figure 6.2 shows the result from such a calculation for the case of an actin gene with $N = 1792$ bp. In this example the melting process starts with the opening of small loops in the 3′UTR (*untranslated region*), while the first peak at 80°C is the dissociation of the whole 3'UTR. The next peak at about 84°C is due to the melting of exons 5 and 6 while exons 3 and 4 melt at a still higher temperature.

Figure 6.3 shows melting domains of various actin genes, all from *H. sapiens*, but from different tissues. Indicated are two regions: the helical (bound) state of the double-stranded DNA at lower temperatures, and the coiled (open) state at higher temperatures. Along the sequence axis the positions of introns are indicated by black lines. In can be seen that the actin genes show a melting preference to the domain boundaries between exons and introns. This preference may indeed have resulted in the insertion of intron sequences on an evolutionary timescale (Carlon et al., 2007).

Nucleosomal signatures of exons and introns

While the physical mechanism of DNA denaturation is already indicative of the involvement of physical effects also in splicing, recent functional genomics results point to yet another picture, related to the presence of nucleosomes. Figure 6.4 sketches two alternative views of splicing, the biochemical view of pre-mRNA ignoring nucleosomes in Figure 6.4a, and a view that involves the presence of nucleosomes, Figure 6.4b. Functional genomics data provide evidence that the nucleosomes are predominantly placed at exons, while introns are largely depleted from them. First of all, there is the circumstantial fact that exon sizes range typically about 140–150 bps, which is pretty much the length of nucleosomal DNA.

In Figure 6.4a the definition of an exon is done at the pre-mRNA level. Spliceosomal and other factors are recruited to the splice sites flanking an exon on the pre-mRNA precursor, which is a single strand of mRNA. Direct and indirect interactions between the 3' and 5' complexes favor exon recognition and splicing and exert selective pressure for a conserved exon length of 140-150 nt. In Figure 6.4b the nucleosomes are preferentially bound to exons, while introns mostly lack nucleosomes. Exons are therefore marked at the DNA level by nucleosome positioning (shown in white), which may act as blocking sites for RNA polymerase II (pol II), helping in the co-transcriptional recruitment of splicing factors to pre-mRNA. As nucleosomes accommodate DNA stretches of approximately 147 bp, their preferential location on exons may act as a selective pressure factor for the conservation in exon length.

On the basis of functional genomics data on *H. sapiens*, *Drosophila melanogaster* and *Caenorhabditis elegans*, Schwartz et al. (2008) provided evidence for the involvement of nucleosomes in splicing. GC content can be shown to be significantly enhanced near the center of constitutive exons. Figure 6.5

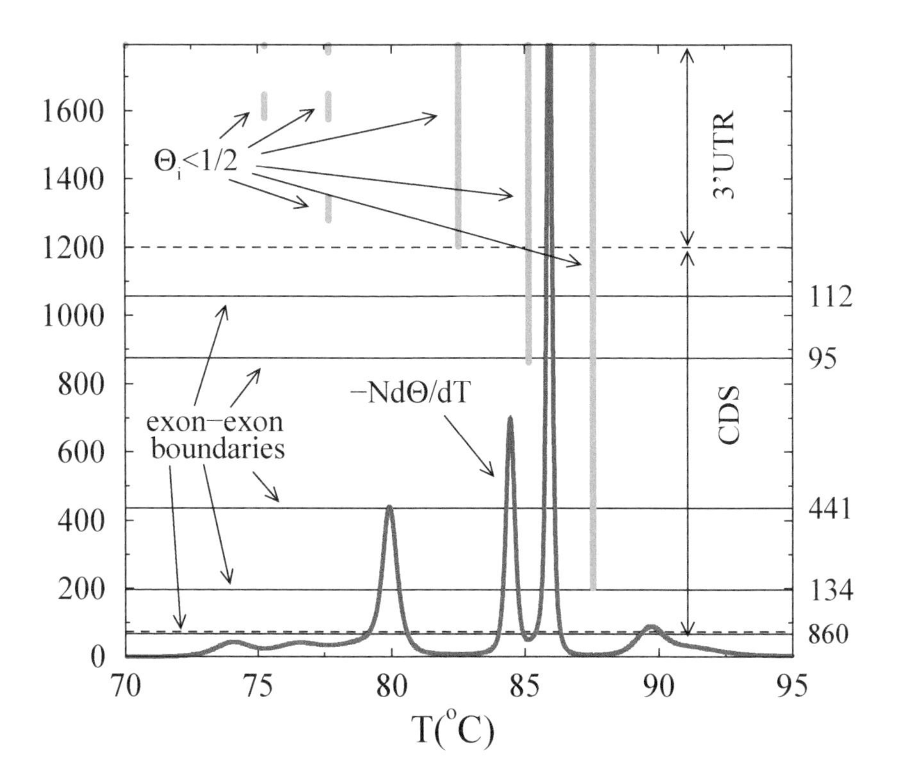

FIGURE 6.2 The differential melting curve and the melting domains for the human β-actin cDNA (NCBI entry code NM_001101). The horizontal axis is temperature; on the vertical axis, $-Nd\theta/dT$ and sequence length. Vertical bars in the graph indicate the regions along the chain for which $\theta_i < 1/2$. Horizontal solid lines are exon exon boundaries and dashed lines are boundaries between the protein CDS and the UTRs. A remarkable overlap between the genomic and thermodynamic domains is observed (Carlon, Malki and Blossey, 2005). Reprinted with permission from the American Physical Society.

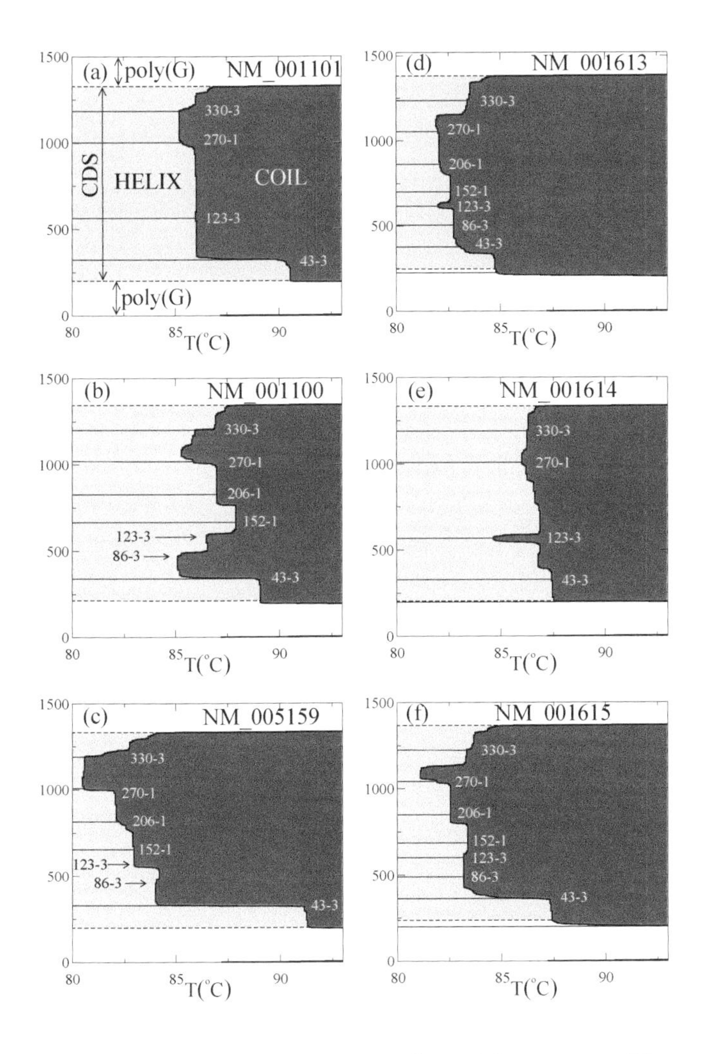

FIGURE 6.3 Melting domains for actins from *H. sapiens*. The numbers refer to different actins from GenBank. The sequences have high similarity to those of other vertebrates, for which almost identical melting patterns are found. In the plots the temperature is the x-axis and the sequence position is on the y-axis. Thick solid lines separate the low-temperature helix state from the high-temperature coiled (open) state. Horizontal dashed lines indicate the boundaries of the coding sequence (CDS) and the solid lines define the intron positions for the given sequence. Arrows point to the intron positions found in homologous sequences (Carlon et al., 2007). Reprinted with permission from the American Physical Society.

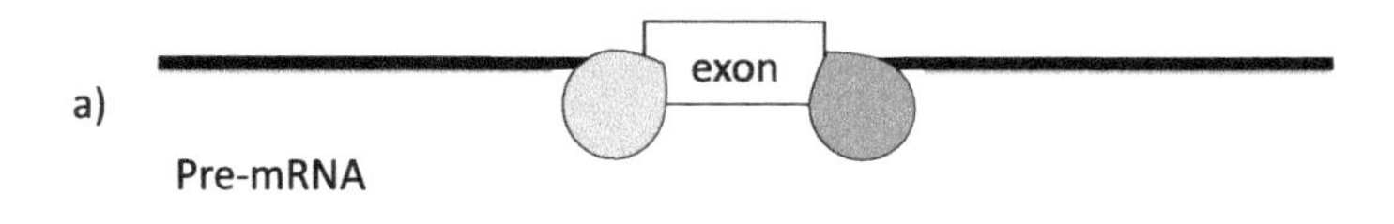

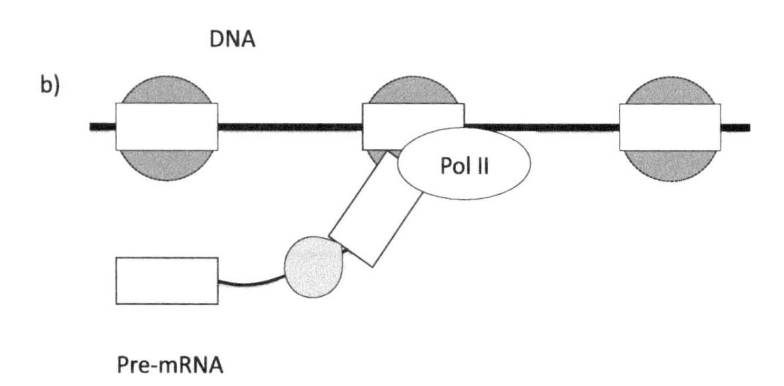

FIGURE 6.4 A schematic drawing comparing the definition of exons in the splicing process, (a), and the involvement of nucleosomes (b). Redrawn and simplified from Kornblihtt et al. (2009).

shows computational predictions for nucleosome occupancy in exons and introns which correlate strongly with GC content. Furthermore, the results by (Schwartz et al. 2008) and also (Tilgner et al. 2009) show that trimethylated histone H3 lysine 36 (H3K36me3) is preferentially enriched in nucleosomal exons, demonstrating yet another correlation between DNA sequence, nucleosome positioning and histone modifications.

DNA repair

In this section we discuss some aspects of DNA damage repair, with a focus on chromatin. Admittedly, this topic deserves a book of its own due to its enormous biochemical complexity, which by far exceeds the possibilities of this book. We therefore right at the outset limit the scope to three questions.

1. What types of DNA damage basically occur that need repair?

2. What is the role of chromatin in this context?

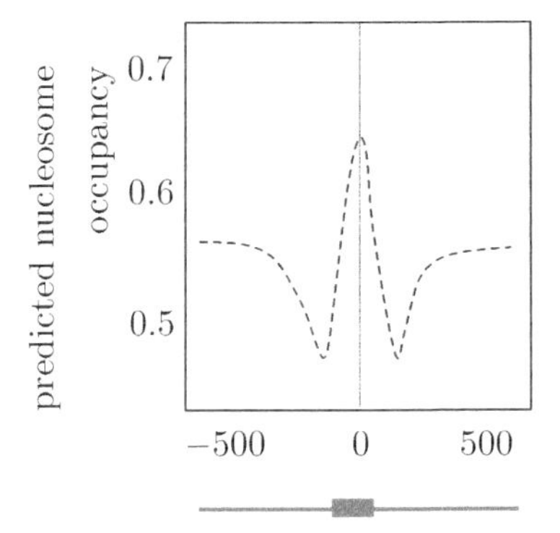

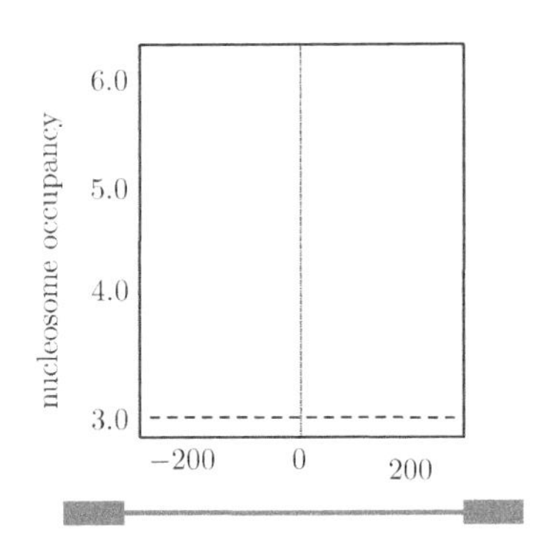

FIGURE 6.5 GC content and computationally predicted nucleosome occupancy in exons. Adapted from Schwartz et al. (2008).

3. How are the mechanisms biochemically principally regulated?

We begin with a short discussion of the types of DNA damage.

DNA repair: single–strand and double–strand defects

Damage to the DNA sequence can be distinguished in several ways. The first type of damage can arise within a properly functioning cell (in some sense it corresponds more to a *wear* than a damage). Such damage can arise from reactive oxygen species, e.g., oxidative deamination which modifies the bases. Secondly, there are environmental effects such as ionizing radiation, most typically UV, X-rays or other. Another way of creating damaged DNA bases is via toxins or mutagenic chemicals.

The most important distinction to be made in DNA damage is whether it occurs on a single strand, in which case the complementary strand is still available as a template for correction, or if one has to deal with a complete double-strand break, which is then obviously more complex to handle for the biochemical repair machinery. We will start with an example of single-strand defect repair based on the *nucleotide-excision repair pathway.*

Kinetic proofreading in DNA repair

In the nucleotide-excision repair pathway a single-strand damage is repaired by the excision of a short segment of DNA around a lesion caused by UV light. Due to UV, nucleotides are modified by the addition of an *adduct* — a chemical which is potentially a cancer agent. UV light leads to the formation of thymine (T) dimers and so-called 6,4-photoproducts. These modifications inhibit the action of polymerase and arrest replication. In order to remove part of the strand an assembly of proteins forms. In the experiments, both in vivo and in vitro, and a model developed by Sancar and collaborators (K.J. Kessler et al., 2007), six repair factors are involved with the names RPA, XPA, XPC, TFIIH, XPG and XPF·ERCC1. These factors eliminate 24–32 nucleotide-long oligomers from the DNA strand at precise locations upstream and downstream relative to the lesion. Figure 6.6 describes the schematics of the interactions, taken from the above paper. One sees first the formation of a pre-initiation complex PIC1 due to the assembly of four of the repair factors. Subsequently, the complex proceeds via a state PIC2 into PIC3. In the upper part of the diagram, the equilibrium assembly of the repair complex is described, covering all ways in which the molecules can in fact intervene. The detailed bookkeeping of these different combinations has been undertaken in order to distinguish between a random assembly and a sequential assembly model, which relies on the idea of a progressive build-up of cooperative interactions. It turns out, however, that, as, in the case of chromatin remodeling discussed before, the key steps involve ATP hydrolysis which lead to the states PIC1-PIC3.

We refrain from a listing of the model equations — the reader interested in the details are referred to the original paper. The reason for this is simply that the number of equations and parameters in the model is already fairly large. It is, however, noticeable that many of the parameters involved have indeed been measured experimentally in the in vitro reconstituted system. Figure 6.7 compares exemplary model results to experimental data for the efficiencies of the observed repair, which show a fair to good agreement with each other.

Obviously ignored in these experiments and the model are chromatin effects. We will include them in the discussion now in two steps, first on the level of a more physical model discussing the effect of lesions on the accessibility of the sequence in the presence of nucleosomes, and then by looking again at the chromatin remodeling dynamics.

The role of nucleosomes in damage recognition

We saw at the beginning of the book how nucleosomes are placed along the DNA fiber. In order to give access to DNA, the histone octamer needs to be displaced from the DNA, and we have seen the passive (by thermal fluctuations) and active (by remodeling) mechanisms that are put in place to achieve this. If a lesion occurs inside a nucleosome, e.g., leading to a single nick, the repair proteins obviously suffer from the difficulty of gaining access to the nucleosome. The question is therefore how the presence of the defect modi-

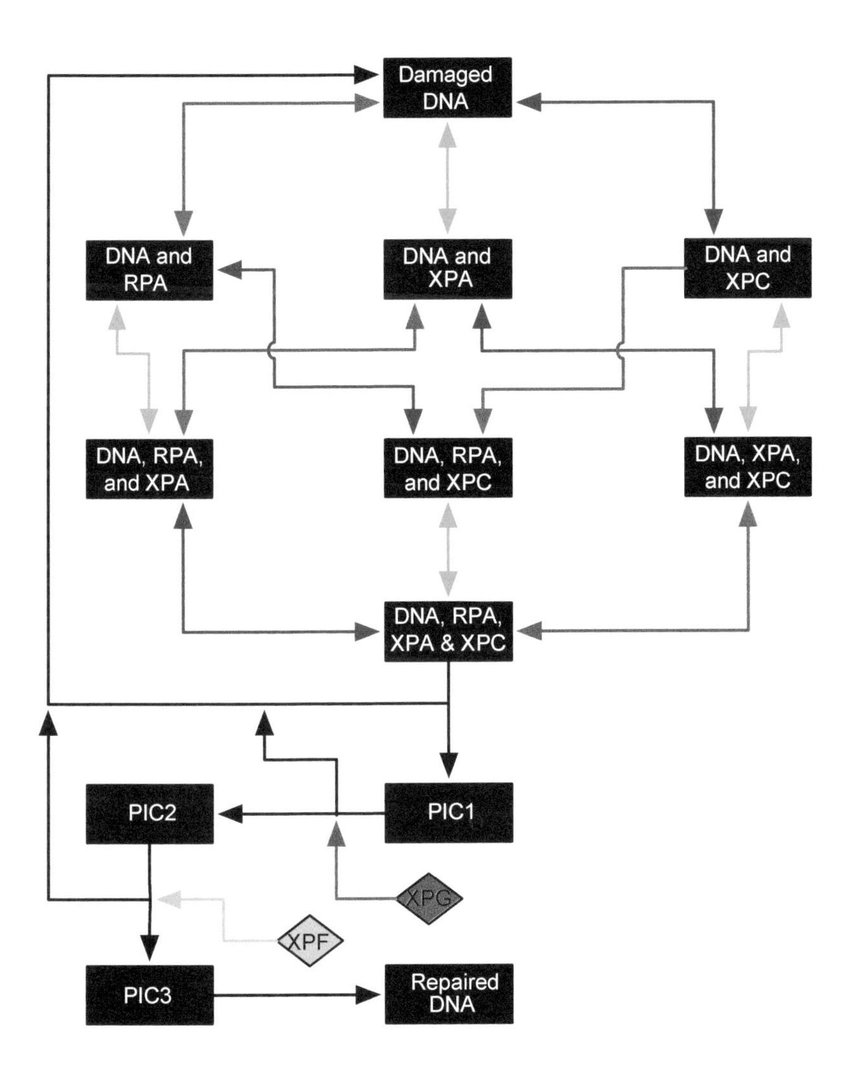

FIGURE 6.6 Schematic of the nucleotide excision repair model by Kessler et al. (2007). Reprinted with permission, © Wiley.

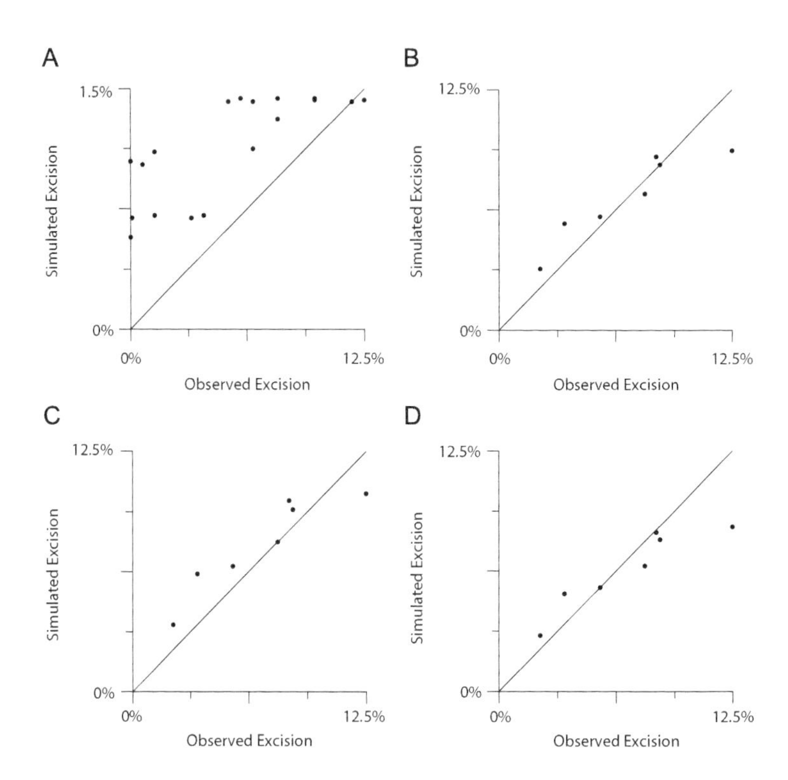

FIGURE 6.7 Excision repair model results compared to experiments. The scatter plots display the amount of DNA excision observed in experiments with ThiT dimers (A) and (6-4) photoproducts (B-D) and the amount excised in a simulation of the model. The maximal observed excision was 1.5% for ThiT dimers and 12.5% for (6-4) photoproducts. Reprinted with permission from Kessler et al. (2007), © Wiley.

fies the dynamics of the nucleosome. We will now study this effect, following LeGresley et al. (2014).

Consider a damage to DNA in the form of a single nick within nucleosomal DNA, say at SHL position 5. The partial wrapping of the nucleosome can be described by a partition function, following Prinsen and Schiessel (2010) who developed this approach for the un-nicked DNA in a study of the work by Widom and collaborators, which we discussed at the beginning of the book in Chapter 1. We denote by the parameter q the energy of wrapping a single base pair on the nucleosomes, times the thermal energy k_BT. The DNA can unwrap from both ends, and one denotes the two ends by the positions x_L and x_R. The length of the wrapped portion of DNA then amounts to $x_R - x_L + 1$, and the partition function thus becomes with the wrapping energy E_w

$$Z = \sum_{x_L=1}^{L} \sum_{x_R=x_L}^{L} e^{-E_w/k_BT} = \sum_{x_L=1}^{L} \sum_{x_R=x_L}^{L} e^{q(x_R-x_L+1)} \approx \frac{e^{q(L+2)}}{q^2} . \tag{6.1}$$

Here, $L = 147$ is the maximal number of base pairs, corresponding to the fully wrapped nucleosome. The probability that a particular nucleotide at position x_B is accessible can then be calculated from summing over all configurations which exclude x_B and normalization with Z.

LeGresley et al. (2014) modified this description by redefining q into an adsorption energy per binding contact, $q_{ad} < 0$, and a bending energy, $q_{el} > 0$. DNA wrapping is then modeled via 14 distinct binding contacts, dividing the nucleosomal DNA into 13 wrapped segments. The centers of these segments are located at the integer positions of the SHLs. The outermost segments are considered unbent. In order to distinguish sequence dependence in the adsorption energy, they are allowed to carry an index n referring to position, $q_{ad,n}$, with n running from -6.5 to $+6.5$. In an analogous fashion, the bending energies are allowed to carry a dependence on the SHL via an index m running from -7 to $+7$. If m_d is the location of a damage site, the partition function for this fixed damage site is given by

$$Z_{fix}(m_d) = \sum_{x_L=-6.5}^{+6.5} \sum_{x_R=x_L}^{+6.5} \exp\left[-\left(\sum_{n=x_L}^{x_R} q_{ad,n} + \sum_{m=x_L+1/2}^{x_R-1/2} q_{el,m}\right)\right] . \tag{6.2}$$

Allowing the variation of the position m_d, one can define

$$Z_{float} = \sum_{m_d=-6}^{+6} Z_{fix}(m_d) , \tag{6.3}$$

so that the probability that the damage is located in segment m_d is given by

$$P(m_d) = Z_{fix}(m_d)/Z_{float} . \tag{6.4}$$

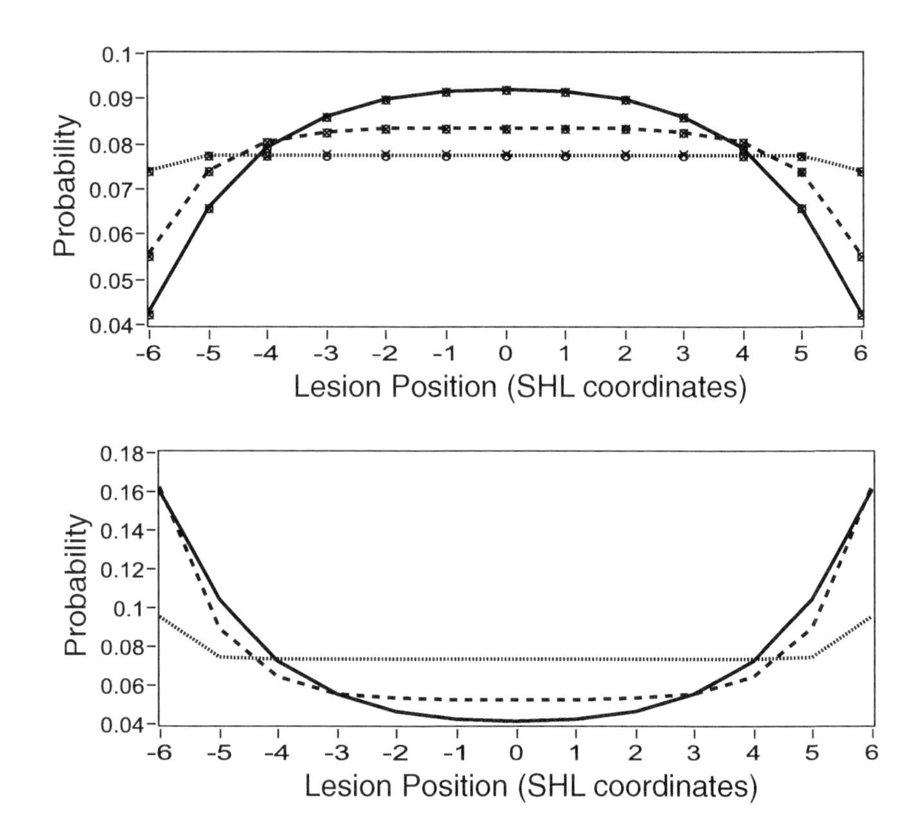

FIGURE 6.8 Positioning of lesions on nucleosomal DNA. Top: flexible lesion with vanishing bending energy for the lesion. For the other segments, $q_{ad} + q_{el} = -0.5\ k_B T$. Solid line $q_{el} = 5\ k_B T$, other lines: higher values; the curves can also be reproduced by $q_{el} + q_{ad} = -1\ k_B T$ (dashed line). Decreasing the sum reduces the positioning effect. Bottom: rigid lesion. Decreasing q_{ad} reduces the positioning effect, from values for $-7.5\ k_B T$ (solid) to dotted, $-10\ k_B T$. The elastic energy of the lesion has been increased with respect to the other segments from $7\ k_B T$ to $9\ k_B T$. Reprinted with permission from LeGresley et al. (2014), © American Physical Society.

We now turn to estimates of the corresponding energies. Within the wormlike-chain model we have $E_{bend} = (\ell_P \ell / 2R^2) k_B T$, with $R = 4.18$ nm, the persistence length $\ell_P = 50$ nm, and an estimate of the wrapping length by $\ell = 3.3$ nm one obtains $E_{bend} \sim 4.7\ k_B T$, resulting in an elastic energy $E_{el} \sim 61\ k_B T$ for the whole nucleosome. For the binding energy one has about $E_{bind} \sim -15\ k_B T$.

The effect of the lesion is essentially to render the corresponding DNA segment straight, which leads to the basic assumption that the corresponding bending energy can be put to zero. One can now start varying the other parameters, and the results from that exercise are shown in Figure 6.8 for the two cases in which the lesion renders the DNA more and less rigid. Overall one finds opposite trends: a flexible lesion will most probably be encountered at the center of the nucleosome, while a rigid lesion will rather propagate outwards.

Regulation of DNA repair in chromatin

In the previous section we saw that the presence of nicks can have an influence on the positioning of nucleosomes under specific assumptions of the involved energetics. This nick-induced breathing effect is, again, one possible physical effect underlying the regulation of DNA repair. Taken together with the biochemical process described before, heavily influenced by ATP-dependent proofreading, we can complete this basic picture now by bringing in the chromatin remodelers.

Ahel et al. (2009) some years ago presented results on the remodeler ALC1 (*Amplified in Liver Cancer 1*), which also goes under the name CHD1L; it is also a member of the SNF2 family of remodelers, which in fact bears some functional similarity to ISWI as its action is not modulated by histone tail modifications. It contains instead a *macro-domain* with which it interacts with poly(ADP-ribose). This polymer is synthesized from nicotinamide adenine nucleotide (NAD^+) by the poly(ADP-ribose) polymerase protein family (PARP). PARP1 and PARP2 are two members of the family that respond to DNA breaks. They carry four domains: first, a DNA-binding domain which is composed of two zinc-finger motifs. Further, the structure contains a so-called caspase-cleaved domain, which plays a role in the inactivation of PARP (an aspect we do not pursue here) and further an auto-modification and a catalytic domain. In the presence of base-pair-excised DNA damage, the DNA-binding domain binds the DNA independently of the other domains. The auto-modification domain is responsible for releasing the protein from DNA after catalysis.

In the experiments reported by Ahel et al. (2009), ALC1 is found to catalyze PARP1-stimulated nucleosome sliding. Depletion or overexpression of ALC1 renders the cells sensitive to DNA damaging agents. PARP1 was found to stimulate the ATPase activity of ALC1 about fourfold. By generating laser-induced damage to DNA, the authors could demonstrate in vivo that ALC1 localized to the positions of DNA breaks. A defective mutant of ALC1 was found to stay bound longer at the break sites. The authors also went on to study the efficiency of DNA repair in the presence of various DNA-damaging chemicals. Therefore, we can conclude from these experiments that in DNA repair we again have to deal with both physical effects affecting the preferred positioning of nicks in nucleosomal DNA, and with a set of tightly controlled biochemical reactions, involving numerous helicase-like complexes which consume ATP for their actions.

Replication: an introduction

DNA replication is the mechanism by which the DNA in the cell is duplicated in the course of the cell cycle in the so-called S-phase; see the box. The duplication of the DNA must be tightly programmed: in the course of one cycle, all DNA needs to be duplicated, and only once. Given the gigantic sizes of the genome, the cell obviously cannot start from one end and then run through; replication must be a parallel process, occurring not only on all chromosomes, but on each and every one at multiple locations. Therefore, the DNA strand is organized in a sequence of replication units (also called *replicons*) in each of which a single starting point or origin of replication (OR) is placed. This OR is, during each cell cycle, only activated once.

> **Cell cycle.** The eukaryotic cell cycle is divided into three main periods called *interphase*, *mitotic phase* or *M phase*, and *cytokinesis*. Interphase is the growth phase, which is itself divided into the G_1-, S- and G_2 phases. The G-phases are growth phases whose ends are checkpoints for the main events in cell division. The S-phase is the phase in which DNA replication occurs. In M-phase the duplicated DNA separates, a process which can again be divided in subphases (called *pro-*, *meta-*, *ana-* and *telo*phases), in which chromatin condenses and orients itself at the mitotic spindle. In cytokinesis, the cells properly divide.

Activation here means that the origin is localizable for the copying machinery, the DNA polymerase, which synthesizes the new strand. This mechanism proceeds from the origin of replication by the opening of *replication forks*: the DNA is locally unwound by a helicase, and thus the sequence becomes accessible. The fork then moves with a replication velocity v along the DNA, and this progression of the replication process at a fork stops when two replicated regions of DNA come into contact. We can therefore image the replication process along the DNA as the opening of replication "bubbles" that will then grow until contact; see Figure 6.9. The different regions also have been given spe-

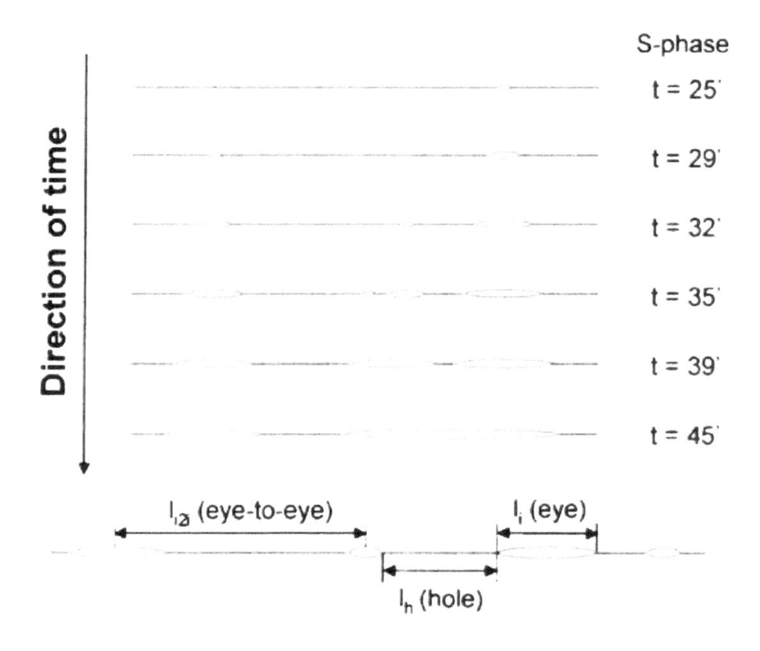

FIGURE 6.9 Replication bubbles and length definitions. Reprinted with permission from Herrick et al. (2002), © Elsevier.

cific names: the replicated region, in which two double strands are present, is sometimes called the "eye", and the region in between the "hole". In the mathematical description we will develop in the following, the respective lengths of these regions will play an important role.

This process immediately makes a physicist think of a one-dimensional nucleation process. Let's take the one-dimensional Ising model for an analogy.

Replication timing: the KMJA model and beyond

Replication can be modeled in analogy to the nucleation problem. In nucleation, the transformation, e.g., from a metastable liquid to a solid phase is initiated by the formation of a critical nucleus which then starts to grow and leads the system into the solid phase. Obviously, there will be more than one nucleus, generally, so the phase transformation will start in several locations fairly simultaneously. These growing domains will then encounter each other and coalesce. We can therefore make the following analogous mappings:

nucleation center (critical nucleus) $\leftrightarrow$ activation of a replication origin domain growth $\leftrightarrow$ bidirectional replication

domain coalescence $\leftrightarrow$ merging of different replicated regions

These mappings can be realized by adopting the KJMA model, originally devised for nucleation in crystal growth in the 1930s; the abbreviation stands for "Kolmogorov–Johnson–Mehl–Avrami" (Herrick et al., 2002; Yang, Rhind and Bechhoefer, 2009). The starting point for the one-dimensional version of the model is the definition of the nucleation rate $I(x,t)$, which is the probability of domain formation per unit length of unreplicated DNA and per unit time, at the DNA position x and at time t. We assume that $I(x,t) = I_0 = const.$ The fraction $f(t)$ of the genome that has been replicated at a given time t during replication (S-)phase is given by the expression

$$f(t) = 1 - e^{-I_0 v t^2} \tag{6.5}$$

where v is the constant replication velocity, which is another model assumption. This expression defines a sigmoidal curve. (Note also that end effects of a finite genome are ignored as well in this model.)

In order to proceed further we need to look at the statistics of replicated regions (eyes), unreplicated regions (holes) and eye-to-eye distances, as introduced before. The corresponding probability distributions can be expressed as functions of time t or of the replicated fraction, f. The distribution of holes of size ℓ at time t is given by

$$\varrho_h(\ell, t) = I_0 e^{-I_0 t \ell} \tag{6.6}$$

from which the mean hole size at time t follows as

$$\ell_h(t) = \frac{1}{I_0 t} . \tag{6.7}$$

The probability distributions of eye sizes are more difficult to find as they may result from different processes: either from a single origin or from multiple origins after mergers. In order to simplify matters one makes another "mean-field" type assumption in which the probability distribution of a given replicated length of DNA is assumed to be independent from its neighbor. One then has for the eye lengths

$$\ell_i(t) = \ell_h(t) \frac{f}{1-f} = \frac{e^{I_0 v t^2} - 1}{I_0 t} \tag{6.8}$$

and the eye-to-eye length

$$\ell_{i2i}(t) = \ell_i(t) + \ell_h(t) = \frac{e^{I_0 v t^2}}{I_0 t} . \tag{6.9}$$

Figure 6.9 illustrates the definitions of these lengths. Figure 6.10 plots the theoretical result in comparison to experimental data obtained from in vitro data of *Xenopus leavis*.

The formulae can be generalized to take into account an arbitrary initiation function $I(t)$. For this one writes

$$f(t) = 1 - e^{-g(t)} \tag{6.10}$$

with

$$g(t) = 2v \int_0^t dt' I(t')(t - t') \tag{6.11}$$

and the equation for the hole lengths then becomes

$$\ell_h(t) = \left[\int_0^t dt' I(t') \right]^{-1} . \tag{6.12}$$

The dependencies of $\ell_i(t)$ and $\ell_{i2i}(t)$ on $\ell_h(t)$ continue to hold. From the experiments, one determines

$$t(f) = \frac{1}{2v} \int_0^f df' \ell_{i2i}(f') = \frac{1}{2v} \int_0^f df' \frac{\ell_h(f')}{1 - f'} \tag{6.13}$$

which one can invert numerically to obtain $f(t)$, as it is a monotone function. One can then calculate all other functions, in particular $I(t)$.

Figure 6.11 plots the replication fraction $f(t)$ in (A) and the nucleation rate $I(t)$ in (B); and in (C) the cumulative initiation density $I_{tot}(t)$.. The maximum in the last quantity allows us to deduce a mean spacing between activated replication origins of 6.3 kb, smaller than the mean eye-to-eye distance of about 14.4 kb. The mean eye-to-eye distance is the inverse of the domain density (active domains per length) from which one can infer the number of active replication forks at each moment during replication.

Replication order: nucleosomes and initiation sites

The above replication model has one essential deficit. The nucleation assumption from the outset obviously relies on the idea that the process starts in a random fashion. In the genome, however, replication starts not from randomly chosen sites but well-defined replication origins. These are, e.g., well-known for budding yeast. The next step to take then is to generalize the KJMA model in such a way that the local initiation rate, $I(x, t)$, can be inferred from these replication origins. Denoting their locations as x_i, one has

$$I(x, t) = \sum_i I_0(x, t)\delta(x - x_i) \tag{6.14}$$

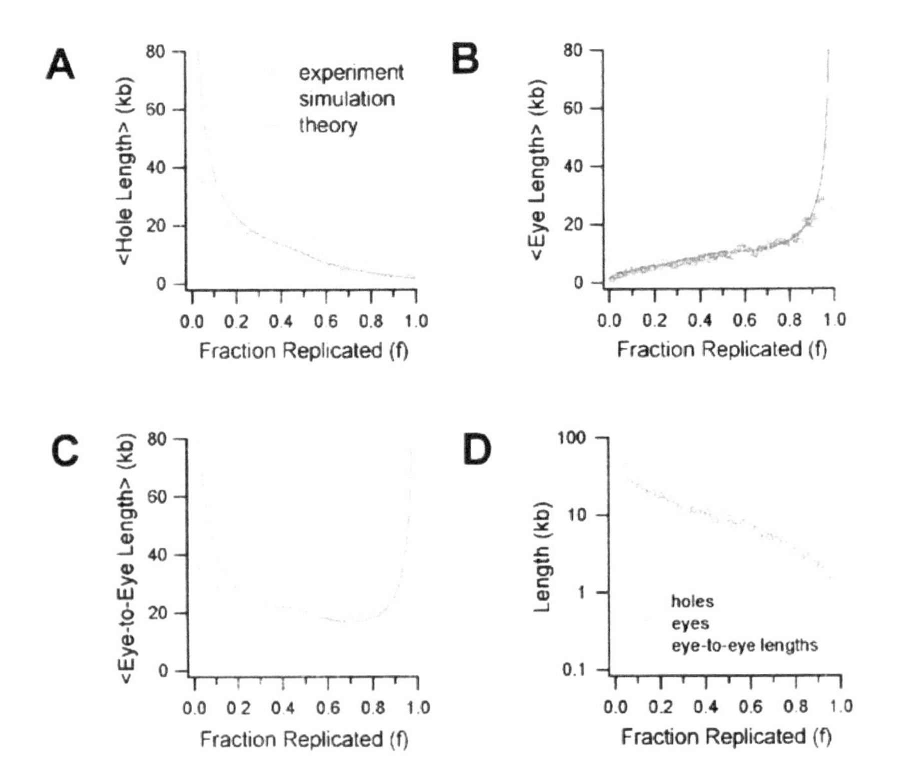

FIGURE 6.10 Mean replication lengths: (A) average hole size, (B) average eye size, (C) average eye-to-eye size, (D) collapsed data. Reprinted with permission from Herrick et al. (2002), © Elsevier.

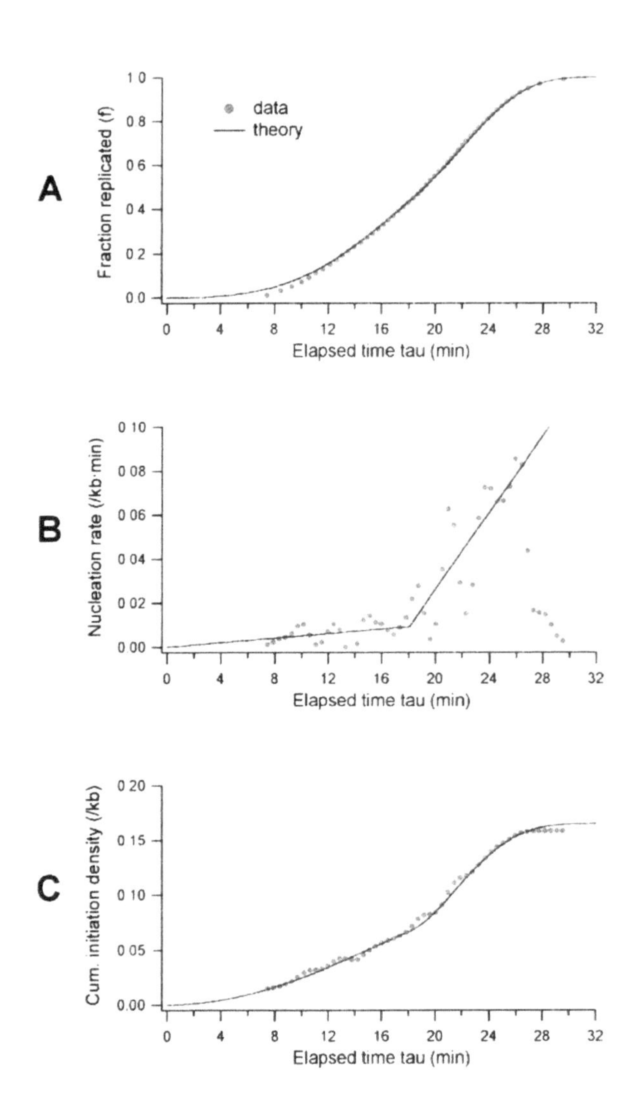

FIGURE 6.11 (A) Fraction of replicated DNA, $f(t)$. The points are derived from the measurements of the replication lengths; the black curve is the analytic fit. (B) Initiation rate $I(t)$. (C) Integrated origin separation $I_{tot}(t)$, which determines the average distance between all replication origins activation until time t. Reprinted with permission from Herrick et al. (2002), © Elsevier.

so that the replication fraction $f(x,t) = 1 - s(x,t)$ can be calculated. For the non-replicated fraction $s(x,t)$ one has

$$s(x,t) = \prod_{\Delta} (1 - I(x',t')\Delta x'\Delta t') = e^{-\int_{\Delta} dx'dt' I(x',t')} \qquad (6.15)$$

where Δ is the shaded region in the space-time diagram shown in Figure 6.12. The origins are assumed to fire independently, and their locus x is unreplicated at a timepoint t if its replication time $t_R(x) > t$. Hence, the probability distribution of replication times follows from $s(x,t) = Prob[t_R(x) \geq t]$: $P(x,t) = -\partial_t s(x,t)$. Because of *passive replication*, the observed distribution $P(x_i,t)$ at an origin i is not equal to the so-called *intrinisic firing time distribution* $\phi(x_i,t)$. The notion of passive replication refers to the fact that when two highly efficient replication origins are placed near each other, only one of them will fire because the initiation of one origin will "passively" replicate the other origin. The intrinsic firing time distribution is given by

$$\phi(x_i,t) = I(x_i,t)e^{-\int_0^t dt' I(x_i,t')}\,. \qquad (6.16)$$

The inverse problem can now be solved by the introduction of "light-cone" coordinates $x_{\pm} = x \mp vt$. In these coordinates, the past light cone of X is understood as $\Delta \equiv \{Y : x_+ \leq y_+, y_- \leq x_-\}$. Hence, taking the logarithm of Eq. (6.15),

$$\int_{x_+ \leq y_+, y_- \leq x_-} dy_+ dy_- I(y_+,y_-) = -\ln s(x_+,x_-)\,. \qquad (6.17)$$

Differentiating this expression with respect to x_+ and x_-, one gets

$$I(x_+,x_-) = \partial_{x_+}\partial_{x_-} \ln s(x_+,x_-)\,. \qquad (6.18)$$

Going finally back to the original coordinates, the last equation can be rewritten as

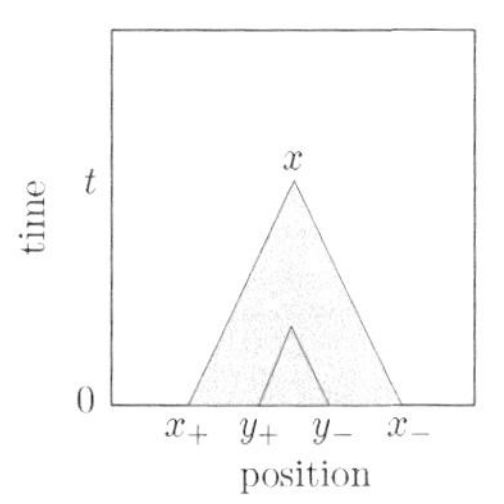

FIGURE 6.12 Space-time diagram after Baker et al. (2012).

$$I(x,t) = -\frac{v}{2}\left(\frac{1}{v^2}\partial_t^2 - \partial_x^2\right) \ln s(x,t)\,. \qquad (6.19)$$

Figure 6.13 displays an exemplary realization of the numerical inversion, undertaken on a 260-kb fragment of yeast chromosome 4 which contains 8 potential replication origins, denoted by O_1 to O_8. The unreplicated fraction $s(x,t)$ was simulated in the top graph, and using the inversion formula the local initiation rate was deduced in the bottom graph.

In the final step we now need to include of the chromatin properties. For this,

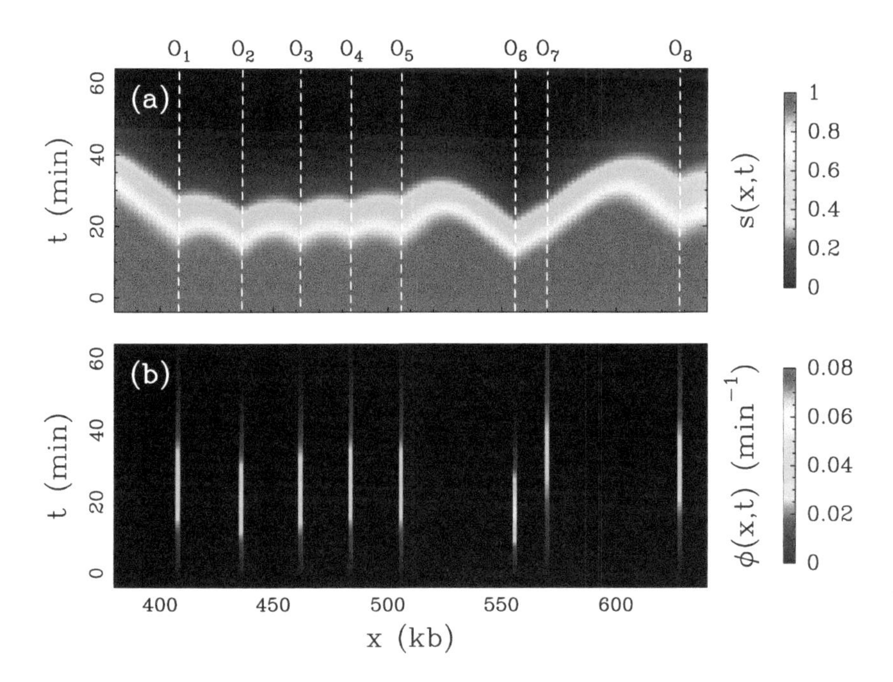

FIGURE 6.13 Calculation of the intrinsic firing time distribution $\phi(x_i, t)$ for a 260-kb fragment of yeast chromosome 4, containing eight potential replication origins. Reprinted with permission from Baker et al. (2012), © American Physical Society.

as in the previous chapters, we resort to functional genomics insights which we however only briefly cite. Cayrou et al. (2015) performed a genome-wide analysis of replication origins in ES mouse cells, and characterized them by both genetic data and chromatin markers. As a result, the replication origins can be grouped into three distinct global classes:

- Class 1 contains isolated, poorly efficient origins with only few epigenetic marks;

- Class 2 is rich in enhancer elements;

- Class 3 has the most efficient replication origins and is associated with open chromatin.

References

Ahel D., Z. Hořejši, N. Wiechens, S.E. Polo, E. Garcia-Wilson, I. Ahel, H. Flynn, M. Skehel, S.C. West, S.P. Jackson, T. Owen-Hughes and S.J. Boulton
Poly(ADP-ribose)-dependent regulation of DNA repair by the chromatin remodeling enzyme ALC1.
Science 325, 1240-1243 (2009)

Baker, A., B. Audit, S.C.-H. Yang, J. Bechhoefer and A. Arneodo
Inferring where and when replication initiates from genome-wide replication timing data.
Phys. Rev. Lett. 108, 268101 (2012)

Carlon E., M.L. Malki and R. Blossey
Exons, introns, and DNA thermodynamics.
Phys. Rev. Lett. 94, 178101 (2005)

Carlon E., A. Dkhissi, M.L. Malki and R. Blossey
Stability domains of actin genes and genomic evolution.
Phys. Rev. E 76, 051916 (2007)

Cayrou C., B. Ballester, I. Pfeiffer, R. Fenouil, P. Coulombe, J.-C. Andrau, J. v. Helden and M. Méchali
The chromatin environment shapes DNA replication origin organization and defines origin classes.
Genome Res. 25, 1873-1885 (2015)

Herrick, J., S. Jun, J. Bechhoefer and A. Bensimon
Kinetic model of DNA replication in eukaryotic organisms.
J. Mol. Biol. 320, 741-750 (2002)

Kessler K.J., W.K. Kaufmann, J.T. Reardon, T.C. Elston and A. Sancar
A mathematical model for human nucleotide excision repair: damage recognition by random order assembly and kinetic proofreading.
J. Theor. Biol. 249, 361-375 (2007)

Kornblihtt A., I.E. Schor, M. Allo and B.J. Blencowe
When chromatin meets splicing.
Nat. Struct. Mol. Biol. 16, 902-903 (2009)

LeGresley, S.E., J. Wilt and M. Antonik
DNA damage may drive nucleosomal reorganization to facilitate damage detection.
Physical Review E 89, 032708 (2014)

Schwartz S., E. Meshorer and G. Ast
Chromatin organization marks exon-intron structure.
Nat. Struct. Mol. Biol. 16, 990-995 (2008)

Tilgner H., C. Nikolaou, S. Althammer, M. Sammeth, M. Beato, J. Valcárel and R. Guigó
Nucleosome positioning as a determinant of exon recognition.
Nat. Struct. Mol. Biol. 16, 996-1001 (2009)

Yang S.C.-H., N. Rhind and J. Bechhoefer
Modeling genome-wide replication kinetics reveals a mechanism for regulation of replication timing.
Mol. Sys. Biol. 6, 404 (2009)

Further reading on replication is

Goldar, A., A. Arneodo, B. Audit, F. Argoul, A. Rappailles, G. Guibaud, N. Petryk, M. Kahli and O. Hyrien
Deciphering DNA replication dynamics in eukaryotic cell populations in relation with their averaged chromatin conformations.
Sci. Reports 6, 22469 (2016)

Further reading on DNA repair: we have evidently only scratched the surface of DNA repair. The following lists some recent review articles on the relationship of chromatin remodeling and DNA repair.

Adar S., J. Hu, J.D. Lieb and A. Sancar
Genome-wide kinetics of DNA excision repair in relation to chromatin state and mutagenesis.
Proc. Natl.. Acad. Sci. 113, E2124-E2133 (2016)

Aydin Ö.Z., W. Vermeulen and H. Lans
ISWI chromatin remodeling complexes in the DNA damage response.
Cell Cycle 13, 3016-3025 (2014)

Gottschalk A.J., G. Timinszky, S.E. Kong, J. Jin, Y. Cai, S.K. Swanson, M.P.

Washburn, L. Florens, A.G. Ladurner, J.W. Conaway and R.C. Conaway
Poly(ADP-ribosyl)ation directs recruitment and activation of an ATP-dependent chromatin remodeler.
Proc. Natl. Acad. Sci. 106, 13770-13774 (2009)

Li G.-M.
New insights and challenges in mismatch repair: getting over the chromatin hurdle.
DNA Repair 19, 48-54 (2014)

Liu B, R.K.H. Yip and Z. Zhou
Chromatin remodeling, DNA damage repair and aging.
Current Genomics 13, 533-547 (2012)

Mjelle R., S.A. Hegre, P.A. Aas, G. Slupphaug, F. Drabløs, P. Sætrom and H.E. Krokan
Cell cycle regulation of human DNA repair and chromatin remodeling genes.
DNA repair 30, 53-67 (2013)

Price B.D. and A.D. D'Andrea
Chromatin remodeling at DNA double-strand breaks.
Cell 152, 1344-1354 (2013)

Rodriguez Y, J.M. Heinz and M.J. Smerdon
Accessing DNA damage in chromatin: preparing the chromatin landscape for base excision repair.
DNA Repair 32, 113-119 (2015)

Tallis M., R. Morra, E. Barkauskaite and I. Ahel
Poly(ADP-ribosyl)ation in regulation of chromatin structure and the DNA damage response.
Chromosoma 123, 79-90 (2014)

Van Attikum H. and S.M. Gasser
ATP-dependent chromatin remodeling and DNA double-strand break repair.
Cell Ccyle 4, 1011-1014 (2005)

CHAPTER 7

Conclusions and Outlook

Much, indeed very much, has been left unsaid in this book. Only one of many omissions in the book is the role of RNA on which we only touched a little bit.

Chromatin has become such an enormous field of research, with scientists from a wide variety of experimental and computational backgrounds, that at present no entirely coherent picture of the field exists. I made this statement in the Preface to this book, but I hope that the reader now appreciates better why this is the case.

In this book I have rather tried to cut a path through the wilderness by looking at chromatin under my particular point of view. I feel that in order to "understand chromatin" one needs to

- properly understand the physical determinants of this polymer-protein complex, and this across multiple spatial and temporal scales;
- determine the biochemical regulatory networks, in particular those involving histone-tail, or epigenetic, modifications;
- understand the role of non-equilibrium, irreversible, ATP-consuming processes, which are key to many specific regulatory functions.

Throughout this book I have tried to highlight in particular this latter aspect, but obviously I cannot offer a comprehensive theory even of this subtopic alone. There is a general difficulty with comprehensive theories in biology; as an illustration, see, e.g., the beautiful paper by Britten and Davidson (1969), in which the authors try to formulate a theory of gene regulation. But the nucleosome was not even discovered at the time.

I like to underscore the relevance of these ATP-dependent processes with recent work by Bruinsma and collaborators (Bruinsma et al., 2014). The authors built a two-fluid model for the viscoelastic hydrodynamics of chromatin including solvent and polymer, and distinguish between equilibrium and non-equilibrium "events", termed "scalar" and "vector". As an example for the latter, ATP-dependent, processes they consider chromatin remodeling. These processes couple in different ways to the hydrodynamic equations: scalar events drive the longitudinal viscoelastic modes, i.e., the polymer relative to the solvent, while vector events drive transverse modes, the overall motion of chromatin (polymer and solvent). They formulate expressions for the flow spectral density (FSD) of the chromatin flow field, finding an expression

$$S(q, \Delta t) = \frac{A(q)}{\Delta t^2} + \frac{B(q)}{\Delta t} \tag{7.1}$$

whereby the wave-vector-dependent coefficients can be fit to experimental data. Although the data they show look qualitatively similar, they exhibit markedly different wave-vector dependencies, which can be ascribed to the two types of events, in the small wave-vector limit, hence on large spatial scales. Figure 7.1 shows the findings, experimental and theoretical, for the case of both ATP-depleted and ATP-non-depleted cells. In this case, for small q, one finds the asymptotic behaviour of the coefficients as

$$A(q) \sim q^{-1.64}\ , \quad B(q) \sim q^{-2.6}\ , \tag{7.2}$$

while for cells not depleted of ATP one observes in this limit,

$$A(q) \sim q^{-1.3}\ , \quad B(q) \sim q^{-3}\ . \tag{7.3}$$

It will be very interesting to link such behaviors to precise biological processes, but nevertheless these results show the relevance of ATP-dependent processes for chromatin as a whole. This book should have made clear that the irreversible, ATP-dependent events are essential as processes that increase specificity of the interactions, alongside the equilibrium processes involving cooperative interactions between molecules.

The story I tell in this book about chromatin remodeling, e.g., suffers from a lack of quantitative data — as do many of the systems under study in chromatin. It is already difficult enough to disentangle the local regulatory networks associated with particular genes — think of the IFN-β gene as an example. Here, we know in great detail the sequence of events, i.e., the build-up of the enhancer, the spreading of the post-translational modifications, etc. But we have no quantitative data, as we do in the case of the λ-phage, that allow us to perfectly characterize the free energies governing these processes, or the non-equilibrium processes involved. While computational approaches might help to rationalize some elements of these findings — as we attempted here — achieving a quantitative theory of even one complex gene is far away.

It is thus for good reason that biologists opt for a different way, which is a functional genomics approach. This approach is quantitative in the sense that

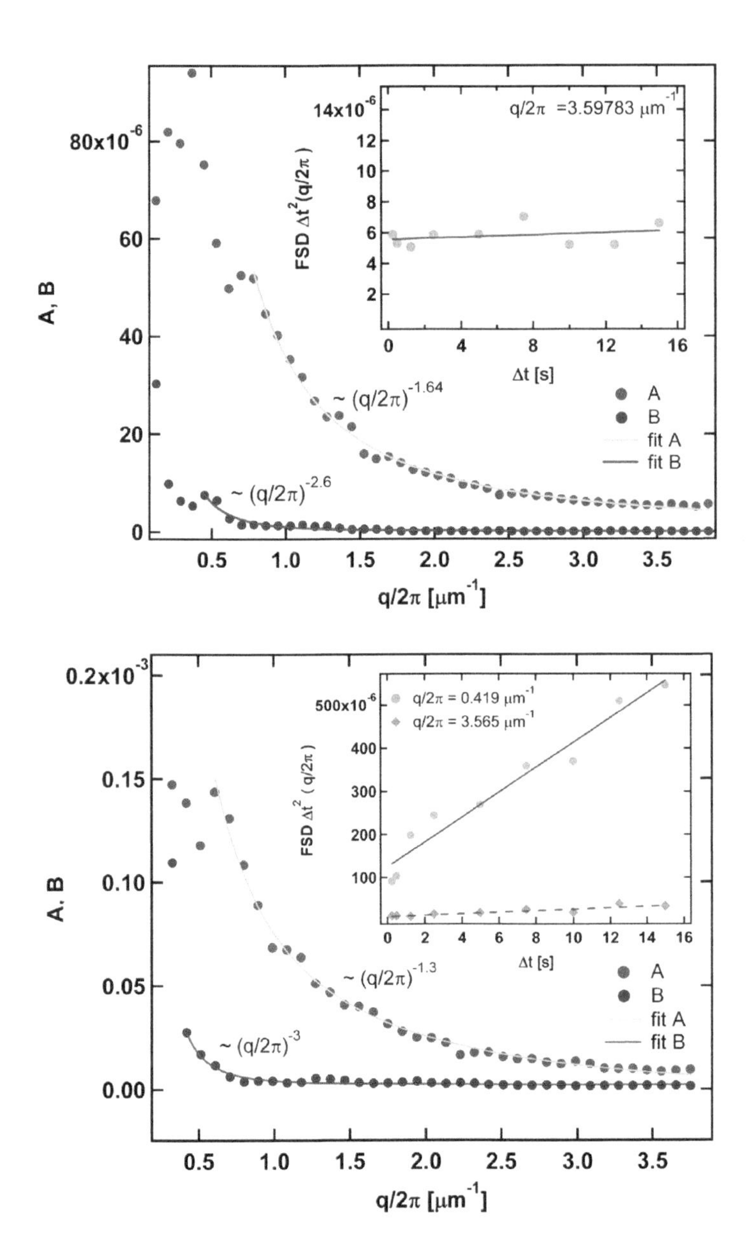

FIGURE 7.1 Flow spectral density (FSD) for chromatin for ATP-depleted cells (top) and non-depleted cells (bottom). Reprinted with permission from Bruinsma et al. (2014), © Elsevier.

it allows us to take a snapshot of "all" processes going on at a specific gene, and to allow for studies of the correlation of specific processes. As an example: if we know how the transcription factors, histone tail modifiers, chromatin remodelers, and epigenetic marks interact, we can fairly well gain an idea of what is going on in the transcription of a gene. In the control of a gene, so many factors intervene that it is only in specific circumstances that one is much more relevant than the other: biological systems are built with enormous redundancy, after all. This is nicely illustrated in the work by Koutroubas, Merika and Thanos (2008) in which they show that by building stronger enhancer elements, steps of the remodeling machinery can be bypassed and still lead to viable cells.

Our discussion should also alert to the fact that since many processes are correlated but not trivially causal, conclusions can be spurious. One can predict nucleosome positions also in part due to the knowledge of transcription factor positions because of their correlation. But nucleosome positions are not causally determined by only these.

I hope that with this book I have helped its readers to find an entry into a complex, puzzling and very alive field of research, with many questions remaining to be tackled in the future.

References

Britten R.J. and E.H. Davidson
Gene regulation for higher cells: a theory.
Science 165, 349-357 (1969)

Bruinsma, R., A.Y. Grosberg, Y. Rabin and A. Zidovska
Chromatin hydrodynamics.
Biophys. J. 106, 1871-1881 (2014)

Koutroubas G., M. Merika and D. Thanos
Bypassing the requirements for epigenetic modifications in gene transcription by increasing enhancer strength.
Mol. Cell. Biol. 28, 926-938 (2008)

Index

For Product Safety Concerns and Information please contact our EU representative GPSR@taylorandfrancis.com Taylor & Francis Verlag GmbH, Kaufingerstraße 24, 80331 München, Germany

T - #0203 - 090625 - C21 - 234/156/10 [12] - CB - 9781498729369 - Gloss Lamination